COSMOS AND MAN

COSMOS AND MAN

A SCIENTIFIC HISTORY

JOHN ENGLEDEW

Algora Publishing
New York

Library of Congress Cataloging-in-Publication Data —

Names: Engledew, John, 1954– author.
Title: Cosmos and man: a scientific history / John Engledew.
Description: New York: Algora Publishing, [2018] | Includes bibliographical
    references and index.
Identifiers: LCCN 2018014131 (print) | LCCN 2018017061 (ebook) | ISBN
    9781628943443 (pdf) | ISBN 9781628943429 (soft cover: alk. paper) | ISBN
    9781628943436 (hard cover: alk. paper)
Subjects: LCSH: Cosmology.
Classification: LCC QB982 (ebook) | LCC QB982.E54 2018 (print) | DDC
    523.1—dc23
LC record available at https://lccn.loc.gov/2018014131

Printed in the United States

Dedicated to Tamara, Devin and Katie

Table of Contents

Preface. Greetings, Sentient Beings

On a personal level I want to express my impressions gathered over an enjoyable lifetime of studying astronomy, history and philosophy (not that I'm finished yet). If I could, I would like to attempt to make a microscopic contribution to the human understanding of the universe.

The majesty of the cosmic nature is motivating in itself, and we will briefly evaluate the universe and mankind's place in it. This book is not a survey of the cosmic and life sciences, but the importance of the unsolved mysteries in cosmology and evolutionary biology are considered. This includes both the enigma of the rise of life on Earth, and its context, and the idea so many of us find attractive and fascinating: the idea that there could be a greater consciousness out there among the stars. Rather, this book aims to sketch out our limited comprehension of it all and the ways in which scientists are now exploring it. Any faults and inaccuracies are entirely my own.

I may take a short flight of fancy along the way. Who was it who said that little of what is worthwhile is ever achieved without dreams?

In his *History of the World*, HG Wells did successfully "refresh and repair his readers' faded and fragmentary conceptions of the great adventure of Mankind." Would that I could do something similar! We can dream that real history is yet to commence and envision some futuristic day when Earth itself will have become a documented legend, rather as we regard the ancient civilizations from today's standpoint. To anyone reading these words from beyond our home world, greetings and every success. You carry the fire of human destiny.

> How strange is the lot of us mortals! Each of us is here for a brief sojourn: For what purpose he knows not, though sometimes he thinks he senses it.

— Albert Einstein

Chapter 1. Man And His Cosmos

"Stars Just Got Bigger"[1]

Human understanding of the cosmos continues to be expanded and refined. We have made powerful progress with our enquiry over the past centuries, yet many new mysteries have been encountered in our odyssey of knowledge. We have come far since the first rational musings of ancient times and the scientific revolutions that led to more recent astounding advances.

The biologist J.B.S. Haldane once remarked that the universe is not only stranger than we imagine but also stranger than we *can* imagine. The cosmos is a dynamically creative place. What does it consist of? How may we comprehend it? How did we get here as both a species and a civilization and above all, what is our place in its grandeur?

To address these questions we commence with a drop of the hard stuff of astronomical science. We shall first consider two of the most enthralling findings of the last few decades, one specific for the size of stars and the other more general about the expansion of the universe. We will intersperse them with the fate of planet Earth, *i.e.*, what will predictably take place here, failing all else.

* * *

In July 2010, a star was discovered in a large cluster with an individual mass some 265 times that of the sun and blazing at 9 million times its brightness. This is greater than thought possible because stars of such sizes were not thought sufficiently stable to exist.

The stellar object, designated R136a1, is a blue hypergiant star within the Large Magellanic Cloud, a detached portion of our own Milky Way galaxy. Specifically, the stellar giant is among the densely populated 30 Doradus Complex or Tarantula Nebula catalogued as NGC 2070. This large emission nebula is 170,000 light years away and a hundred times the size of the Orion nebula, the glowing gas cloud that is the naked-eye example of its type. A light year is, of course, the distance a ray of light travels in a year, propagating at 186,000 miles a second. This is equivalent to nearly six million million miles and is a necessary unit of distance for even the closest stars.

The Tarantula Nebula is powered by the ultraviolet radiation emitted from some young, hot and massive stars embedded within its expanse. The French astronomer Lacille first recognized its nebular nature in the mid 18th century, and in 1987 the closest supernova observed since the invention of the telescope occurred in an outlying region. As stellar associations go, it is a very busy area of activity. In all, the super cluster R136 contains an estimated 100,000 members, including many Wolf-Rayet type stars, an advanced stage of development when they lose their hydrogen envelopes to leave their helium cores exposed.

Weighing in as the new stellar heavyweight champion, R136a1 is now the most massive and luminous individual star known. Its surface temperature is an estimated 25,000 K, which is the level in degrees Kelvin above absolute zero equivalent to -273° C. (In expressing degrees Kelvin, the symbol ° is unnecessary.)

Its luminosity is estimated at 8,700,000 times that of the sun. At 50 million miles in diameter, it could be an unresolved pair among four other hypergiants or the result of major stellar mergers among close binaries. In this concentrated region there are twelve other members of comparative mass (or M_{suns} as this is expressed).

When light is spectrally analyzed from stars, it gives information on their chemical and physical conditions. Their constitutions and temperatures are further revealed by the encoded information revealed in spectral lines. Looking back, the philosophers Comte and Hegel made utterances that the makeup of celestial bodies would forever be beyond man's knowledge, shortly before the first development of spectroscopy. It has long since been a prime tool of astrophysics, including the estimates of stellar distances and speeds by spectroscopic parallax.

Estimates hold that the R136a1 has shed a quantity of matter equal to 50_{suns} in the last one million years and will likely explode as a hypernova in the cosmological near future. Normally, the upper limit for the amount of nuclear material a star can possibly sustain and remain stable is an estimated 100–120 $M_{suns.}$

Like other stars close to this Eddington limit, it dispenses much of its super heated material through the continuous gaseous outflow of stellar winds. Hence, such a limit is not a strict constraint on their luminosities. It is merely the hypothetical point where the gravitational force inward equals the continuous radiation force outward to maintain stability. Beyond this, a star is unable to hold itself together due to the loss resulting from the outflow and in theory would lose so much material as to push itself apart. R136a1 probably commenced with some 320 M_{suns} of material.

Generally, more massive stars go through their life cycle at faster paces and the vast majority of them are far more modest. Our local star has a greater mass than Earth by a factor of nearly 333,000 and is quite average in all other statistics of composition, age, mass, temperature and luminosity. Its diameter of 865,000 miles is 100 times that of Earth and 1,000 times that of Jupiter, the largest planet in the solar system. For our first piece of perspective, the sun is an ordinary star among billions.

What of the superstars in the Milky Way?

There are at least twenty known stars with radii exceeding a thousand times that of the sun, with fully accurate measurements are often uncertain. This is due to their extended atmospheres, pulsating variability and obscuration by dust. Often their true sizes remain undefined because much of their radiation lies in the infrared rather than the visible portions of the spectrum.

In radius the largest known star remains the stellar object catalogued as VY Canis Majoris, a red hypergiant approximately 2000 solar radii in size, at a distance of nearly 5000 light years. This prodigious quantity of star material is spread over a larger volume than R136a1. VY Canis Majoris burns at a more familiar 3,000 K, cooler than the surface of the sun's effective 5778 K. Refined measurements suggest values of 1420 +/- 120 the solar radii.

The red supergiant and pulsating variable star UY Scuti, lying 9,500 light years away, has an estimated 1708 +/- 192 solar radii. Its diameter of some 1.5 billion miles is larger than Saturn's orbit around the sun. It is also 5 billion times the volume. Elsewhere, the so called Pistol Star radiates as much energy in 20 seconds as the sun does in a year and the Eta Carinae stellar system stably outshines the sun by factor of 5 million.

Considering the long-term actions of stars, major byproducts include the heavier elements that are forged within them by nucleosynthesis. In turn these enrich the interstellar medium in elements heavier than hydrogen and helium such as oxygen, nitrogen and carbon. It is an important set of processes, because at great length this provides the chemical building blocks that presumably give rise to life within hospitable planetary biospheres.

What has unfolded here on Earth allows us to imagine that it must surely occur elsewhere? Unfortunately, we are extrapolating from a single example, but for a point of departure it is remarkable that the human race is here at all to contemplate such things. That alone gives us pause and we will examine this more closely. It also reveals the limits of present knowledge because, after all, life is only definitively known to exist on Earth. We will now address what will ultimately happen to our planet itself. Is no surprise that this is linked to the long-term future of the sun.

The sun comprises 99.86% of the mass of its entire entourage of planets, moons, asteroids and comets we call the solar system, the full range of bodies in the gravitational grasp of our local star. Our world is third in order of the four small, rocky and innermost bodies we call Mercury, Venus, Earth and Mars. As my sister noticed in our childhood, Earth is the only planet not named after a god. Much further and in larger orbits of configuration are the four gas giant worlds of Jupiter, Saturn, Uranus and Neptune. The latter two are better described as ice giants and Pluto as the primary dwarf planet and least distant member of the far-flung Kuiper belt or Edgeworth-Kuiper belt.

Assuming no other cataclysm strikes, the final outcome for Earth relates to how the sun is fundamentally changing as it progresses along its own path of evolution. In time it will become a red giant like the familiar Arcturus or Gamma Crucis, which at 89 light years is the closest example.

The greater red supergiants like Antares and Betelgeuse have radii greater than the sun by a factor of 1000 and are conspicuous over hundreds of light years. In appearance, the starry sky gives the impression of a greater commonality of red giants because they are so luminous over distances. They actually comprise a mere 1% of the broad stellar population.

In the full course of time, the core of the sun will undergo change. Its busy nuclear fusion has consumed a mere 0.03 % of its total reserves over an existence of 4.6 billion years and this is at the ongoing rate of 4 million tons of hydrogen (H) converted into helium (He) every second. (Reading about this last statistic and its release of energy first drew me to astronomy as a child. That and the sheer scales of space and time associated.) Our local star was likely slightly larger in its earlier career, probably a few hundred degrees hotter and 6% greater in radius when initially formed.

In the huge nuclear power piles of stars the outward pressure of fusion is typically balanced by the pull of gravity to produce dynamic stability. Further on when hydrogen becomes exhausted the innermost core gravitationally contracts, directing the material to enter the areas where conditions of temperature and pressure allow further hot atomic processes to commence. The reaction rates and release of energy accelerates to include helium fusion and the outer layers of the star booms to the red giant phase. With our sun this will occur in some 3.8 billion years' time or something less than twice its current early middle age.

Beyond that the helium in the shell becomes depleted and the star contracts until a new shell He reaches inward and ignites, discharging most of its outer parts in a visible blast of superheated gas known as a planetary nebula. This is something of a misnomer. The Ring nebula M57 is an example of such conspicuous rings of material surrounding their progenitor star that are relatively brief in phase. We comprehend enough of stellar types and evolution to say that planetary nebulae are relatively brief in phase and that the sun is unlikely to become a supernova, being of insufficient mass.

Astrophysics is able to make a positive prediction about the long term fate of planet Earth. Later in its progress as a star, the sun will engulf the bodies occupying the inner solar system.

We go all the way round the sun in a year and the earth lies in an average 93-million-mile wide orbit, the definition of the Astronomical Unit used in planetary astronomy. When the sun begins to enlarge, the closer zones including our orbit will be absorbed into its superhot outer atmosphere if not its active body.

It is presently about 4.6 billion years old and the process will entail a gradual brightening over the forthcoming 6.5 billion years as the sun leaves what is called the main sequence of the Hertzsprung Russell diagram of stellar types. During the next period of about 20 million years it will progress to the asymptotic giant branch to reach the red giant phase of maximum luminosity and size.

Not that the earth will be immediately swallowed whole in a short period. In expanding and discharging perhaps one third of its mass, the sun's gravitational field will slowly become more extended over a newly bloated area. The great source of attraction holding the worlds in their courses could effectively loosen its grip, possibly allowing the inner planets to shift outward and assume wider paths.

Other variables include the slowing rotation of the sun, its own tidal bulging, and the collective influence of everything in its gravitational grasp. But a sun growing over 5,000 times in luminosity and 200 times greater in diameter than present is assured.

Hence, the planets may slip further out than their present positions. The more distant members are less likely to be vaporized by such massive heating although higher temperatures will certainly come to prevail out there. The depths of the sun's empire will finally rise from the ambient freeze pervading since its formation. Hypothetically, the processes giving rise to organic activity and possibly life might be kick started in the previously cold depths of the eons.

Superheating will certainly end all life here, causing a boiling away of the atmosphere and oceans and a parching of the surface. By this remote future epoch the earth will be spinning more slowly and attended by a more distant moon as the enormous heat vaporizes them both.

Whilst this nemesis for life is inescapable, there is a possible whole alternative to this loosening of the sun's gravitational grasp. Instead, the spreading hot outer atmosphere and far stronger solar wind of ejected particles could tidally slow the inner planets. The closer worlds could be moved incrementally nearer as their orbits slim down by the increasing friction in the interplanetary medium, allowing them to be consumed more speedily. Either case will seal their fate.

The discussion is therefore one of solar gravitation allowing their paths to slip outward versus deceleration drawing them gradually inward. In the fullness of time, everything in the inner solar system will be destroyed or completely distorted by the expansion of the sun as it changes and a much higher temperature gradient ensues. For now, our beloved Sol is about one third of its way through its active life and we have more pressing problems than its long-term status, many unrelated to long-term stellar lifecycles.

The constancy of the sun is vital to us, but super flares have been observed in other sun-like stars and we can only hope ours does not undergo some sort of upheaval, however transient.

There is nothing we could do if we were faced with only a small change in its behavior or the strength or type of its energy output. Our technology is woefully insufficient to save us from a solar cataclysm on even a minor scale. We comfort ourselves that such a thing is unlikely to occur for now, but in the cosmic sciences, lessons in humility are a constant leitmotif. We will review a range of cosmic doom scenarios in Chapter 3.

By the evidence of tree rings, the sun did produce a super flare *c.* AD 775, and more generally it is a slightly variable star with a cycle taking 8–12 years. A longer-term variability related to its output of energy is possibly related but probably not the sole cause of the great and small ice ages or other climatological effects in Earth history. It has possibly increased in luminosity by 1% over the last million years. To us, the sun has always been a special body, occasionally revered as a deity as such. As the original wellspring and sustainer of life there was some pragmatic point in sun worship.

Putting this in proportion, there are 108 broadly solar type stars within a sphere of 16 parsecs. (A parallax second is equal to 3.26 light years.) Incidentally, we hold it coincidental that our closest stellar neighbor is so similar to the sun, both formally designated G2V in class, spectral type and luminosity.

The sun may well have first formed amidst a cluster, but the same Alpha Centauri system is significantly older and demonstrably not a lost associate. Stars are often born in groups like binaries or clusters, remaining so throughout their careers, but these are merely two independent stars just over four light years apart. This is typical spacing in this backwater of the local Milky

Way we call the Galaxy. Alas, any original siblings of the sun at formation are now entirely lost to identification, and scientific papers on the matter prove inconclusive. Considering their proper motions in space over time, few of the neighboring stars were originally formed where they currently are.

For better orientation, our address is modest to the point of obscure. We reside within the Orion Spur of one galactic arm of the same name and other arms comprised of stars surround a central bulge. The center of the Galaxy forms a flattened spheroid about 8,000 light years across its poles and 10,000 over the diameter of its equator. The inner regions were first to form in our typical barred spiral galaxy containing some 100,000 million regular stars and untold numbers of lesser stellar and planetary bodies. Nowadays we must emphasize that these are merely the visible and non-dark matter components.

For long-term futurists there is another guarantee, this one on a higher level involving the Local Group of galaxies. Four billion years from now the Milky Way is expected to merge with the Andromeda Galaxy.

There are a few such galactic mergers on show like the Antennae Galaxies that were separate entities 1.2 billion years ago. In action such coalescences bring on distended shapes and gigantic gravitational effects yet the situation with the component stars is like the interaction of swarms of bees. With the spacings of stars, there are few actual collisions.

Whilst our immediate galactic neighbor is still 2½ million light years away, the name Milkomeda is prepared for the single island universe that will result. Nothing like good long-term planning! Assuming we still exist, man could be successfully established across the local star systems and will no doubt have been physiologically changed as a species.

So much more waits to be expounded in cosmology, the study of the large-scale structure, history and evolution of the universe. Contemporary debates include the rôle of dark energy in its ongoing expansion and the veracity of M or multiple universe theories. We will not resolve such issues anytime soon, but during our time together we will be taking an ambitious glance. The universe is an extremely inventive place. As an inspiration both analytically and aesthetically, I recommend the former to gain some balanced appreciation.

Perhaps the grandest mystery of all is how common life may be, especially the question of whether creatures on planets orbiting other stars have achieved intelligence and applied it to paths of cultural and technological development. Speculation has been lavished on the matter to the extent of inventing exobiology, the hypothetical study of life beyond Earth. It is a subject in sore need of an object and a theme we will be exploring at length.

The question could be practically refined to ask: How many advanced alien civilizations lie within a 1,000 light years of the sun? Here is a simple but profound enquiry if ever there was one and a matter integral to any realistic sense of place. There are approximately 1,000,000 stars within such a radius. I suggest the answer would be straightforward if only we could get to grips with some real information. How many stars are orbited by worlds inhabited by sentient beings at such immediate stellar distances? For that matter, how common or rare is life of any kind expressed as a percentage of parent stars?

We simply do not know. Refer to the Drake Equation, still the best tool for any quantitative estimate of planets and life arising and developing into technological civilizations. If only one star in a million is host to intelligent life, then the figure emerges of 100,000 groups of little green men in our Galaxy alone. These form the bases of SETI and CETI projects, the search for and attempt to communicate with extraterrestrial intelligence.

At the heart of the subject, Fermi's Paradox still reigns: Where is everybody? Absence of evidence is never evidence of absence, and by any inference of statistics the logical conclusion is that other intelligent creatures must exist somewhere. By this reasoning, we should not expect that mankind is the only such advanced creature. Admittedly, the search has been disappointing so far, meaning that we indefinitely endure that eerie silence while waiting for any direct evidence of alien civilizations. Perhaps we are going about communication with others in the wrong way, or alternatively, we could be on the cusp of a major breakthrough. Whatever the rarity or commonality of advanced life (and I think the former applies), we simply cannot assume we are the only flicker of rational consciousness or pulse of cerebral enlightenment that ever emerged since the Big Bang.

The second most significant news is more general to cosmology, consisting in the identification of a general acceleration in the outer parts of the observed universe as a major surprise to us all. It was the dramatic result of the studies of very distant Type 1 supernovae by their redshift and luminosity distances.

Up until the late 1990s, many scientists thought we were approaching an approximately confident understanding of the layout of the cosmos, called the grand design. Some plumbing of the greatest scales of time and space seemed within our grasp. Standard Big Bang cosmology and relativity are notable achievements but with the discovery of actual acceleration in the expansion a revised approach becomes necessary. It is termed the extravagant universe.

Cosmologically, the great architect of gravitation may have contributed some influence of deceleration for some 6–7 billion years following the Big Bang. Before dark energy began to take over to repower and drive further and faster expansion. This remains speculative but I think future research will support the case. That the further observed reaches are actually speeding up reveals the important part played by this potent vacuum energy.

Earlier in the story of our discovery, a major step occurred in the mid 1960s with the serendipitous discovery of cosmic microwave radiation. As engineers, Penzias and Wilson were mapping the radio sky, engaged in a far more modest project for Bell Telephone Laboratories; they hardly anticipated winning the Nobel Prize for one of the greatest scientific findings ever. Similar circumstances had attended Karl Jansky's first discovery of radio waves from the Milky Way and sun in 1931. Latterly, the source of CMR proves to be a recorded glimpse of the cosmos at a stage of development of recombination during the event termed *decoupling*.

We take our first step in to cosmology to cite that this occurred some 380,000 years after the Big Bang as the general density fell and the universe had become transparent to radiation. With photons newly free to travel undisturbed along wholly unlimited free paths they became cooled by the great expansion. Previously the plasma state of the cosmos had been opaque. These photons are the very ones responsible for the observed CMR as background and from our vantage point became a received omni-directional remnant signal at this stage following the earlier super explosive expansion.

The "shortly afterward" of some 380,000 years later is clearly the blink of an eye in cosmic time. By several corroborated methods we confidently gauge that 13.7 billion years have elapsed since the Big Bang or when the universe was 1/36,000 of its current age. This is how and when we think everything began, an explosive formative event that included the very emergence of space and time. The CMR has a thermal black body spectrum at a temperature of 2.725 K and the glow is very nearly uniform in all directions *i.e.*, isotropic.

The COBE (Cosmic Background Explorer), WMAP (Wilkinson Microwave Aniostropy Probe) and Planck satellites have provided informative and detailed mappings of the very early state of the cosmos. The improving images represent the moments when specific hydrogen atoms formed from a great and energetic soup of electrons and protons and the universe was no longer that opaque to radiation.

A snapshot when temperatures fell to some 454,000 K reveals a template for the future development from a fundamentally younger age. We deduce that the tiny residual variations show a very specific pattern. Therefore, we have a veritable baby picture of early creation and major hints for its forward development are nested in those images.

Gamow and others had predicted a radiation relic and it is acknowledged that the Russian radioastronomer T. A. Shmaonov briefly observed it in the mid 1950s. There had been several other anticipations of its existence, such as Dicke, Peebles, *et al.* at Princeton estimating a microwave background radiation with a temperature as low as 3.5 K. When Dicke got the call from Bell Telephone Laboratories concerning the Bell team's discovery, the very thing his team was working on, he's quoted as saying, "Boys, we've been scooped."

It is now half a century since the discovery of CMR and it must be integrated into any successful model of the early universe. Portions of everyday TV static and the snowy interference on blank screens originate from CMR. It is still my favorite show.

A little more on the history of modern cosmology is in order.

Lemaitre, De Sitter and Friedmann and others had independently hypothesized about a highly condensed original state before the concept of a great expansion took hold. The latter proposed a universe whose radius increased over time as one solution to Einstein's equations as early as 1922. It was apparent that the cosmos should be either expanding or contracting and a decade earlier, Vesto Slipher had obtained the radial velocity, actually a spectral blueshift of approach for the Andromeda Galaxy. The other few examined galaxies showed a redshift of recession and such a motion towards the Galaxy in the case of Andromeda proves a notable exception. At the time no results had been generalized indicating a general expansion.

Prior to any real concept of a "Big Bang," there had been scientific discussion on the big picture, culminating in the Shapley-Curtis debate that took place in 1920. The first astronomer held that those uncertain nebulae were localized regions within the Milky Way and comprised the entire cosmos. The other hypothesis placed island universes as wholly external to our own Galaxy.

It proves that there are many other independent galaxies beyond our own as Shapley asserted but the other view and was more accurate about the scale of the Milky Way and the lowly position of our sun in galactic geography. Shortly after, the epochal telescopic work of Edwin Hubble better interpreted the spectra of galaxies to discover their expansion. It was a great step forward in understanding the universe.

In 1931, Lemaître published an article in *Nature* employing the novel term "primeval atom" or "cosmic egg" as the idea of a single and smaller point of origin. If the universe is expanding over time, there must have been a point in the past when everything was much more compact. Being a man of the cloth as well as a scientist may have worked against the acceptance of Monsignor Lemaître's ideas. He must be seen as a pioneer of good ideas in cosmology, once informing Einstein that his mathematics was good but his physics abominable.

Fred Hoyle originally used the phrase "big bang" as a disparaging critique of a concept he strongly opposed. It was in a radio broadcast in 1949 that he inadvertently supplied the name for posterity. Ironically, it gave ammunition to his opponents whilst his group proposed "continuous creation" as a rival hypothesis. The theory had its day, but part of its downfall was CMR, whose incidence was not easily accounted for in this same and opposing "steady state" hypothesis.

He did substantially better with nucleosynthesis, the processes that create new atomic nuclei from pre-existing protons and neutrons in the inner workings of stars. This is the accepted mechanism of the production of elements beyond H and He in their processes of stellar nuclear fusion. We speak of the metallicity of stars (Z) as a fraction of all other elements present in stars and those with relatively high abundances of carbon (C), nitrogen (N), oxygen (O) and neon (Ne) are referred to as "metal rich." Note that these are astrophysical terms and not metals as defined in chemistry or their positions on the periodic table of the elements.

Earlier in the 20th century Einstein had triumphed with the Special and General theories of relativity but had entirely gone along with the concept of a static universe. Nothing had previously suggested it could be otherwise. Among his sterling work he had invoked a cosmological constant as a mathematical term to balance its full collapse, and in his equations it was designed to "hold back" gravity. A grand gravitational collapse of the universe by its own attraction is not a happening thing.

With Hubble's work at Mt. Wilson, the actual conditions of galaxies rushing apart and something of their faster rates over increasing distances were decisively demonstrated by photography and spectroscopy. The further galaxies are shifting at higher speed, and there is a direct relationship of greater distances to their higher rates of recession. For many years the precise relationship of Hubble's constant of distance to speed of recession was so quantitatively uncertain as to be the most inconstant constant in physics.

Einstein's self-confessed "greatest blunder" of proposing a cosmological constant to prevent such an implosion is currently making a most useful comeback; now that dark energy is found to conduct such a crucial part in the accelerating expansion of the universe. A non-zero cosmological constant has been necessarily revamped to describe what is going on, something originally designed to maintain an assuredly static state of affairs. Both the origin of and speculative possibilities for the outcome of the cosmos will be addressed later.

Moving up to date, we can itemize the more exacting results from the European Space Agency's Planck mission as follows. The conclusion, derived from sixteen months of mapping the cosmic microwave background and announced in March 2013, declared that the universe is slightly older than WMAP's figure, giving us 13.798 +/- 0.037 billion years of age.

Also the fraction of dark energy comprising the cosmos is slightly lowered from the previous estimate of 71.4% to 69.2%. Dark matter exists as a proportion of 25.9% and actual baryonic matter a mere 4.9%.

As for the background noise, the temperature is equal to 2.725 48 +/- 0.00057 K.

The Hubble constant, the relation of recessional velocity to distance on galactic scales has hopefully been finalized too. The rate of increase based on local standard candles rises to 73.8 +/- 2.4 km/second/megaparsec.

For an observation heralding a new era, the gravitational wave ripple from two black holes merging over three billion years ago was confirmed to have occurred for a fifth of a second in September 2015.

Dawn Of The Space Age

The Space Age was inaugurated by the first man-made satellite placed in Earth orbit. It commenced when the former USSR successfully launched Sputnik in October 1957. Geopolitics and spaceflight have changed so much since then that it is worth reiterating the first steps. Propelling a craft aloft fulfilled dreams as old as mankind and we will describe the selective highlights of the story so far.

Sputnik consisted of a metal sphere 23 inches in diameter. It travelled at 18,000 mph at a height of 359 miles. With batteries powering a radio signal through four external antennae, the first satellite circled the world every 96 minutes and remained aloft for 92 days over an orbital trek of 43.5 million miles before reentering the atmosphere and burning up. In the moment, the political implications were as imposing as the technological achievement.

However much we liked Ike or jived to Elvis, this demonstration of the advanced state of Soviet science showed that the USSR was more than a rival socio-economic model; with their capability in launch boosters there was consternation that they could hurl a missile at us more efficiently than we could blow them up. It hinted further at an acrimonious "missile gap" in the wake of the hard and bitter peace following the Second World War.

President Eisenhower was not surprised at the Soviet contribution to the International Geophysical Year but had badly misjudged domestic reactions to the "Sputnik effect." What were our rivals capable of now? The following month, Sputnik 2 carried a canine cosmonaut into orbit, the first creature from Earth to enter space.

In response, the Americans attempted to launch a Vanguard satellite, but it resulted in an explosion on the pad. Seen as a display of technological inferiority, newspapers called it "flopnik." Using a Juno rocket, the US Army finally got Explorer 1 into orbit in January 1958 but the next two American efforts similarly met failure.

There were real risks that the Cold War could hot up. President Truman had previously taken a full military initiative against the spread of communism in Korea and foresaw similar developments in Vietnam. A U.S. spy plane was downed over Russian territory in May 1960, creating an embarrassment for the next administration and as a clash of ideologies, pessimists held a Third World War inevitable. In his departing words, Eisenhower warned of an industrial-military complex and earlier in his administration he had told the gentlemen of the press:

> Every gun that is made, every warship launched, every rocket fired, signifies in the final sense, a theft from those who hunger and are not fed, those who are cold and not clothed. This world in arms is not spending money alone. It is spending the sweat of its laborers, the genius of its scientists, the hopes of its children ... this is not a way of life at all, in any true sense. Under the cloud of threatening war, it is humanity hanging from a cross of iron.

Shall we eulogize Eisenhower as a leader who had learned the horrendous cost of war? He certainly had first-hand experience on the highest level. On any account, events were now moving swiftly in the new frontier of space. By April 1961 the succeeding president, Kennedy, was to endure the news of a Russian cosmonaut in orbit and the Bay of Pigs fiasco, which failed to take back Cuba by force. He was a mere three months into office. The torch being passed to a new generation of Americans looked more like a baptism of fire and the later Cuban missile crisis was probably the closest we came to nuclear apocalypse. Kennedy once said he was an idealist without illusions.

* * *

Rockets were nothing new on limited scales. Chinese troops had used them to target arrows in the 13th century, having invented gunpowder itself three centuries earlier in the pursuit of potions allowing immortality. There is a fanciful tale of enough rockets being attached to a chair to have once blasted a Mandarin astronaut into the sky.

The British army probably reached the technological limit of the age during the War of 1812, shooting off enough rockets to be evoked in the Star Spangled Banner. Beyond that, their serious uses were few beyond propulsion for harpoons and rescue lines.

Modern rocketry to hurl payloads of explosives aloft was born from the research and development left over from the fall of Nazi Germany. Before Von Braun and his team were brought here to work for the fledgling U.S. space program, their wartime careers built and launched over 3,000 V2 rockets. These newly designed liquid fueled ballistic missiles were successors to the V1 or buzz bomb that had been the first of a new type of pulsejet powered cruise weapon.

The trajectories of V2s rose some 50 miles and touched over 3,000 mph before reigning destruction by high explosive and the kinetic energy of impact. They were the only "super weapon" to operationally rise from the drawing boards of the doomed Third Reich. Seeing a successful launch ascending 52½ miles, an artillery officer named Dornberger saw fit to declare that "this third day of October 1942 is the first of a new era in transportation, that of space travel..."

The Führer originally saw rocketry as no more than a means for firing long-range artillery shells at prohibitively higher cost. Not deeming them worthy of intensified development, he only enthused about mass production of the "vengeance" weapons and other schemes when it was too late. With a range of about 200 miles, the aim of the V2 was sufficient to strike targets like London and other cities, and there was no defense against them. Eisenhower remarked that if they had been available a mere six months earlier, there could have been a different outcome in Europe.

Von Braun had been a member of the Verein für Raumschiffahrt in his youth. This Society for Spacetravel was an authentic amateur research group that built a few modest large rockets and dreamed of spaceflight before the rise of Hitler. Similar circumstances surrounded the Messerschmitt Me 262, the first operational jet ever built. They were the fastest aircraft yet and the rocket-powered Messerschmitt Me 163 Komet was even swifter, but they could not turn back the terrible tide engulfing Germany's defeat. As for being unequivocally first into space, a V2 was lofted to an altitude of 108 miles in June 1944.

This surpasses the boundary of the atmosphere as set by the Fédération Aéronautique Internationale, lying at the altitude of 100 kms (about 62 miles). This Karman line is based on the altitude where the atmosphere becomes too thin to support aerodynamic flight because a vehicle would have to travel faster than orbital velocity to derive aerodynamic lift to support itself (centrifugal force excepted).

As early theorists of rocket-driven spaceflight, Von Braun, Oberth and Ley must be accorded true pioneers, as should Goddard as an inventor and engineer. Tsiolkovsky, the pioneering Russian scientist (working in obscurity as a rural school teacher) had conceived of multi-staged rocket boosters and an orbiting space station as early as 1903.

A camera mounted on a V2 captured by the Americans provided the first pictures of a small section of the earth from space in 1946. It was a remarkable footage, shot from an altitude of 65 miles reached in an ascent of mere minutes. Missile development and boosters to launch scientific payloads were now inextricably linked in both the US and USSR. A race to build and test and produce ever-greater weapons gripped both superpowers.

As for manned vehicles before ballistic or orbital flights by either nation, thirteen American X-15 rocket powered aircraft ascended over 50 miles in a program commencing in mid 1959. The X-15's set a new speed record of Mach 6.72 and as one test pilot a certain pilot named Neil Armstrong flew 7 of the 199 flights. A decade later he and Aldrin became first to set foot on the moon.

Consider a celestial body 2,000 miles in diameter in orbit about the earth that is moving at just under 1 mile a second at an average distance of 239,000 miles. Plus, a minimum speed of 25,000 miles an hour is essential to set out from Earth and no longer be held by its major field of gravity by ever falling back. This imposing 7 miles a second for escape velocity under controlled propulsion then requires navigational guidance in space, itself a new science. These are the fundamental logistics of any attempted lunar mission and in

January 1959 the unmanned Russian Luna 1 crossed cislunar space to successfully reach the vicinity of the moon.

Luna 1 was also the first rocket ignition in Earth orbit as it accelerated
out from the approximate 18,000 mph to aim as close as possible to its objective. Intended as an impacter it actually passed within 4,000 miles of the
moon before entering an independent heliocentric orbit somewhere between
Earth and Mars where it remains to this day. Suitably, it designers renamed
it *Mechta* (dream) and the "first cosmic ship."

For scientific discovery its onboard instruments discovered the solar
wind and indicated no detectable lunar magnetic field.

Years later it was revealed that three previous lunar attempts by the
Russians had suffered launch failures. Two months later the American Pioneer 4 managed to pass within 36,000 miles of the moon but in a landmark
achievement the second successful probe of the Russian series made actual
impact. It turns out that Luna 2 had been preceded by another effort failing
to achieve Earth orbit, but in September 1959 it was the very first manmade
vehicle to contact another cosmic body, as did its booster.

Within months, Luna 3 provided the first fuzzy images of the moon's far
side. It certainly left an indelible impression on me. As a child, I was awe-
struck at those indistinct pictures in the newspapers of its unknown far
hemisphere. It was a momentous sight. The new images from space were
something truly new in human experience and my family and friends have
not had any rest since.

(N.B. On being first to escape the earth's gravitational field with anything
other than a rocket-borne payload, the extravagant American astronomer
Zwicky received permission to place explosive pellets about an Aerobee
rocket that was sent aloft twelve days after Sputnik. Two of these small objects attained escape velocity.)

Why should "rocket science" be infamous for being so tough to comprehend? The most important concept is Newton's third law, which states that
for every action there is an equal and opposite reaction. Rocket propulsion is
still the only known form functioning in the vacuum of space, rendering warp
factor fives and quantum gravity drives, *etc.*, as inspiring fiction, for now.

In the new decade, both nations successfully orbited more satellites. Two
dogs rode aboard Sputnik 6, recovered alive this time and the U.S. sent a

chimpanzee on a suborbital hop, drawing some criticism as a worthy objective. Were we really spending the taxpayers' money to put monkeys into
space because we are afraid what the Russians are doing up there? In a further feat of engineering the Soviets launched Venera 1 towards the planet Venus in February 1961. Two months later the Russians took a step significant
in the story of all-time human achievement, putting the first man in space.

Gagarin's single orbit of the earth lasted 1 hour 48 minutes and finished
with the cosmonaut ejecting as planned in the final phase of descent to parachute safely back to Russia. Somewhat off course, his Vostok vehicle landed
with a bump. It is now on permanent display in a museum near Moscow
and I touched it myself on a visit. Following his triumph, the handsome and
charming cosmonaut was given huge publicity and international tours that
were a far cry from the abject State secrecy that characterized the former
USSR. He made quite an impression and among many honors he was awarded the prestigious title Hero of the Soviet Union.

Confronting the rumor challenging whether Yuri Gagarin was the first
man in space, there is now far better access to the records and commentar-
ies on the earliest Russian missions. No longer dogged with terse official re-
ports always released *after* events and with no mention of launch or staging
failures, it is conclusive that he was indeed the first human space traveler.
Also rejected are vague tales of four cosmonauts killed in separate suborbital
flights going awry, and that one Vladimir Ilyushin had previously completed
an orbit but was badly injured on return. Another version has him landing off
course in China where he was supposedly detained for a year. Conclusively,
Ilyushin was a decorated test pilot but never trained as a cosmonaut.

The following May, Venera 1 probably reached within 80,000 miles of
Venus, although contact had been lost after only seven days. Its sensors had
showed the solar wind to be fully extant in near space but again we now
know that a previous Venus project had failed to leave Earth orbit.

Remarkable steps have since been taken and entire new visions have
been unfurled by space probes. We know our neighbors so much better now
as real worlds.

In a way, Mercury is the least often explored of the inner planets, having been called on only twice. These were the pioneering Mariner 10 that
also flew past Venus before making our first encounter with the innermost

planet in the early 1970s. (The eleventh and twelfth Mariner probes were al-located to the future Voyager program.) However, the more recent Mercury Messenger achieved 100% surface mapping over four years in orbit before finishing with a planned crash. A joint ESA–JAXA project to Mercury is cur-rently projected.

With Venus the record stands at 38 successful or semi successful mis-sions and about 12 failures including the first attempt to land there. We put this Soviet Venera 4 in the partial success category because it met tempera-tures and pressures ending the telemetry and active mission at an altitude of about 16 miles. Venera 4 reached the surface of Venus in pieces. It could not have landed intact.

There are active contemporary rovings on the surface of Mars with an-other vehicle specifically being built for 2020. We understand much more about the red planet by so many visits after the first success of Mariner 4 in 1964. NASA's historical log lists a total of 44 attempts by four nations in all with a failure rate occasionally called the "Mars curse." Far more successful-ly, the Mars Reconnaissance Orbiter is still in action after a full decade, one of a collective six satellites presently circling Mars plus two operational craft on the surface. To date, a collective nine flybys or orbital missions have been made to Jupiter and four to Saturn and a single encounter with Uranus and Neptune was achieved with Voyager 2.

Many of the satellites of the outer planets have now been viewed from close range. Over a full 20 year mission the Cassini vehicle became the first craft ever to enter Saturn orbit and part of its mission was to release the Huy-gens probe to successfully soft land on the moon Titan. Cassini completed about 300 orbits of Saturn over 13 years and was today deliberately de-orbit-ed into the atmospheres of the ringed planet. Another long awaited step in our explorations was reaching Pluto in July 2015 when the New Horizons vehicle arrived safely after a 9½-year voyage.

In an overview of unmanned spaceflight, it completed an initial survey of them all. This sounds rather trite for the voyages of exploration undertaken and the resources so brilliantly applied but we do have the most impressive record of visiting every major and numerous minor member of the solar sys-tem between 1959 and 2016.

Put it this way: Imagine no space programs had taken place to date then some inspired scientist or ambitious politician proposed to do from scratch over the next six decades what has been actually achieved in that time. He would be laughed off the lectern or stump.

Whilst the Apollo project to put men on the moon was initiated as an adjunct of the Cold War, the race to the moon accrued a bounty of scientific knowledge apart from the political prestige of winning it. Originally devised as one way of beating the Russians in a new arena of space, Kennedy laid down the goal in memorable speeches. At the time, America had conducted a mere four single man Mercury missions with the first two 15-minute sub-orbital flights. It was a total of less than 11 man-hours in space by American astronauts at the time. After the trauma of Kennedy's assassination, it was fortunate that the succeeding LBJ continued to share enthusiasm for man's greatest adventure.

Humans walking and driving on the moon and the deployment of scientific instruments on its surface remains the best achievement in manned spaceflight so far. Can Apollo be done justice in a mere paragraph?

The timeframe is from October 1968–December 1972. After Apollo 7's eleven-day inaugural flight of an American three-man vehicle in Earth orbit came Apollo 8's first manned voyage to the moon. It was a "gutsy mission" (Chaikin) and by far the most exciting one yet over Christmas 1968. It proved that translunar injection, the voyage out, lunar orbit insertion, ten orbits and the first trans earth injection for the return journey to reentry and recovery could be done with a manned vehicle. (Note that two Russian unmanned Zond vehicles with one containing life had successfully circumnavigated the moon and been returned shortly before.) Apollo 9's crew then tested the lunar module landing craft around the earth. Apollo 10 was the "shakedown" mission whose lunar module descended within 50,000 feet of the surface. The came Apollo 11, that one small step for (a) man, one giant leap for mankind as Armstrong stepped onto the moon, followed by Aldrin. The plaque on the LM reads, "Here men from planet Earth first set foot upon the moon July 1969, A.D. We came in peace for all Mankind." Apollo 12 was another success despite camera problems and the full record includes twelve men setting foot there in six landings and a total of nine missions involving twenty-four astronauts. (Three individuals got to go twice.) Naturally, the story includes the highest speeds ever attained in human space flight and the deepest voyages into space ever made by astronauts. Specifically, these are Apollo 10s TEI and Apollo 13's single loop around the moon at a height of 103 miles respectively. Apollo 13 was a "successful failure" (Nixon) when the landing was scrubbed, turning the mission into a cliffhanger to get the crew back alive. After the moonwalks of Apollo 14, EVA's were better equipped with lunar rover vehicles during 15 and 16. In the final mission, Apollo 17's two astronauts were on the surface for three days, driving over 14 miles in three EVA's. In the annals of exploration, comparisons to Columbus and the voyages of discovery leap to mind, as do the expeditions reaching the poles and highest mountains. The feat of walking on the moon may be equated on the same level as creatures leaving the seas of Earth to adapt to the land millions of years ago. It is that important to the human story. The Hubble Space telescope recently provided views of the six landing sites where the tiny images of the lunar module descent stages and other equipment stand as monuments to the magnificent achievement of Apollo. In hard figures twelve astronauts over six landing missions spent an accumulated 80 hrs 34 mins walking on the surface of the moon over a series of stays lasting a collective 12 days 11 hours 28 mins.

It is worth sketching the story leading to Apollo. The Russian Vostok and American Mercury missions had been the first one-man capsules during the early 1960s. There was a joke that the astronaut did not so much go on board as put it on as equipment. He was "the man in the can." In high political profile each program flew six missions with Vostoks 3 and 4 in orbit at the same time and Vostok 6 carrying Tereshkova, the first woman in space.

Most of the "firsts" went to the Russians including Voshkod 1, a vehicle carrying three men that also set a new altitude record of 208 miles. With Voskhod 2, Leanov, one of the two cosmonauts exited the craft in orbit and over twelve minutes performed the world's first spacewalk.

In the moment the Soviets were passing these milestones before NASA had finalized their two-man Gemini spacecraft, giving the impression that the three man Voskhod series was fully operational and the Soviets were far ahead. However, the Gemini flights probably put the Americans in the lead by late 1966 including dockings, more spacewalks and a new altitude record of 850 miles with Gemini 11, America's seventeenth manned excursion. It was valuable experience to realistically reach the moon before the decade was out.

It is now clear that the two Voskhods that flew were designed for only *two* crewmen and it was for propaganda that three cosmonauts were ordered to fly the first mission. The craft itself was mostly a larger version of Vostok and there was no room for wearing spacesuits on Voskhod 1, an inordinate risk.

As for the moon race, one opinion is that that the Russians were content to watch the Americans fail and then go to the moon themselves in due time. It is difficult to fully assess this. Instead, political decisions, the failures of their giant N1 boosters, a poorer plan of progress and the death of Korolev, their chief designer caused them to abandon the whole project by the early 1970s. In this time the Americans had flown astronauts to the moon nine times. In fretting about the Soviet capability to get to the moon first, it seemed unlikely in 1968 or to this day that no Russian has ever ventured beyond Earth orbit. Kennedy had privately contemplated a collaborative moon mission with the Soviets but there was never any real proposal.

Much could be said of capitalistic enterprise and openness triumphing over totalitarian authority. More specifically the Americans pursued a better step-by-step approach to design, build and launch and far more transparency of intent. Mere days before Apollo 11, the unmanned Luna 15 had been launched in an attempt to sample and return but it crashed there whilst Armstrong and Aldrin were taking the first moonwalk. It was their last chance to upstage the U.S. by getting geological samples back first. Following further American successes they seemed to deny there had been a competition and proposed cooperation in space.

After this joint Apollo-Soyuz Test Project, the Russian Salyut was the first space station followed by the American Skylab in the 1970s. From 1986–2001 Mir included eleven rendezvous by American Space Shuttles and since then the International Space Station has taken up the mantle of a permanent human presence in space. Looking back to the halcyon days of Apollo it is a matter of regret that three further missions to the moon were cancelled, their Saturn V boosters in preparation if not lunar modules already built. Extended missions and visiting the lunar farside had been slated.

Finally, in our thumbnail history of astronautics, let us consider the direct human costs. The darkest days of Apollo were in January 1967, with the deaths of three American astronauts in a ground test resulting in a fire in the command module. In its wake, the consensus holds that the major redesigns forced on NASA's spacecraft prevented a subsequent disaster in flight.

One lesson learned the hardest way was to eliminate a pure oxygen environment onboard. O_2 supports combustion and in a ground test an electrical spark produced a deadly inferno aboard Apollo 1. With the funerals of Chafee, White and Grissom some suggested we should bury the whole program. White was the first American space walker and Grissom first to fly twice in space. Both nations had other fatalities in training.

Later that year, the Russian cosmonaut Komorov tragically became the first man to lose his life in an active space mission. Like Voskhod 1 in which he had flown, it was another political decision to launch the new Soyuz vehicle although it was not ready. Among a maze of flaws, the final blow came from entangled parachutes causing the vehicle to strike the ground at high speed. After successful flights in that program, the demise of the three-man crew of Soyuz 11 in June 1971 reveals that the cosmonauts were asphyxiated by depressurization as they prepared to reenter, giving them the sad distinction of being the only humans to perish outside the earth's atmosphere. They were the sole crew to dock and successfully enter Salyut, and the Soyuz vehicle bearing their bodies landed successfully.

Appropriately, there are mementos and a roll of honor by name to them all on the surface of the moon, including one for Gagarin, who died in an air crash in 1968.

Up to January 1986 NASA had never lost an astronaut in flight but two Space Shuttle missions were to claim their entire seven person crews. For the record, this first reusable Space Transportation System consisted of an original four and an added fifth vehicle that flew 135 missions and 355 astronauts over two decades. But in full view of the world, there was a catastrophic explosion on the tenth launch of the Challenger and the twenty-fifth mission, grounding the program for over two years. Seventeen years later, with STS 107, the returning Columbia broke up on reentry having suffered damage on launch.

We honor the memory of our fallen heroes. All progress comes at a price, and every one of them would want us to continue the quest.

Of late, there has been much talk of a crewed expedition to Mars. A diverse "Mars underground" of feasibility studies has existed since the literary flights of Jules Verne and Tsiolkovsky's calculations and diagrams. They were independently composed on paper for entertainment and scientific speculation. The French writer was far more influential with his spaceships of the imagination voyaging to the moon in 1865 but one quote from the Russian schoolteacher who became the father of astronautics is definitive:

> "The Earth is the cradle of the human mind, but we cannot live in the cradle forever."

Nowadays, there are no more shrill negations of the possibility of humans flying in space, as its early proponents had to put up with. This kind of talk had driven Goddard and his experiments with rocketry into seclusion. However, serious logistical challenges in life support, propulsion and other areas remain to be solved before we can realistically set foot on Mars. At its very closest it is still 140 times further than the moon and a viable vehicle hardly exists as a blueprint for what would be an expedition lasting several years at achievable speeds. There lies the crux.

Encouragingly, we are not short of volunteers to go or even stay indefinitely. Nothing is impossible if it breaks no fundamental laws of nature, including long term missions to Mars, constructing space elevators to orbital altitudes or astronauts hitching rides on asteroids. We must never fall victim to failures of either nerve or imagination in the march of science and pursuit of knowledge.

As Jules Verne said: "Science, my lad, is made up of mistakes. But they are mistakes which it is useful to make because they lead little by little to the truth."

With the retirement of the space shuttles, there is currently a hiatus for the U.S. in terms of any major booster capability. Note that Soyuz vehicles are used to reach and return from the ISS and there are several new developments from private space industry in ascent to orbit. Observing that no one has been beyond earth orbit since the final Apollo flight, it is high time this torch was passed. In the fullness of time, an American Vanguard remains the oldest satellite still circling the earth.

The Europa Clipper and an ESA mission to Jupiter's moons are approved next plans and it is enthralling about the prospects of Breakthrough Listen and Breakthrough Starshot. In the latter case the proponents are serious about reaching the closest star system Alpha Centauri with nano spacecraft, small enough to hold in the hand. With the discovery of an exoplanet for Proxima Centauri, they certainly have somewhere to go. Let us be confident that the technical challenges can be overcome and this may come to fruition in the coming century.

A Better Understanding Of The Solar System

In a perfect world, any space agency would find the resources to swell the boundaries of knowledge by both unmanned missions and human astronauts. Imagine a happy and peaceable human world united in the drive for space exploration where high adventure and the frontiers of science were the greater goals.

Wouldn't that be nice in a sane and progressive global society motivated by such ideals and free of poverty, disease and conflict? Or that giant space stations orbited the earth and long-term colonization projects were top on a fully cooperative international agenda? There are many stirring eulogies that our destiny lays beyond Earth and unlike budgets dreams have no limits.

Taking the lesson from Tsiolkovsky, in the distant future we may be primarily remembered for taking the first steps beyond our planet of origin. It might define the present human era among our positive achievements. Let us review just some of the programs underway in unmanned spaceflight and interpret what has been learned in planetary science.

I shall defer broad descriptions of the planets and the probes to them to other sources. Our understanding is so much better since the Mariners, Pioneers, Voyagers and others went out there. They make for fascinating contemporary tales of discovery and nice pictures adorning the textbooks. Those images of Uranus and Neptune close up are from our sole Voyager encounter.

What of the lesser bodies of the solar system in the story so far? What is their context for a better overall understanding? What are the limits of the solar system? Bear with me on some technical details as we turn attention to some of the minor league players.

For encounters in space the current score is thirteen asteroids and eight comets with one failed mission named Contour or the Comet Nucleus Tour of 2002. This was curtailed when contact was lost on leaving Earth orbit and the vehicle probably disintegrated. You can't win them all. What were our successful forays and what new knowledge has been gathered?

In 1993, the swift passing of asteroid 951 Gaspra was a notable sideshow for the Jupiter bound Galileo spacecraft because it revealed the appearance of a minor planet from a range of 2,000 miles. An authentic close up picture of an asteroid at last! We had waited for years.

Later on its journey, Galileo imaged 243 Ida over six hours at a close point of 1,490 miles. It was found to possess a moon of its own that we named Dactyl, an asteroid with a satellite or the existence of binary asteroids being a surprise. In early 2001 we completed orbital maneuvers and successfully set the Near Earth Asteroid Rendezvous vehicle down on 433 Eros. This was not just a close flyby but a man-made probe taking up residence on the surface of an asteroid. It is a permanent memorial to our developing abilities as a space faring species.

The first mission to a comet was the International Cometary Explorer, passing through the tail of comet 21P/Giacobini-Zinner) in September 1985. It also made rendezvous with the most famous Halley's comet (1P/Halley), as did two Russian Vega stations via a passing of Venus where descent units had been successfully deployed to the planet's surface. The best images of the Halley nucleus came from the ESA's Giotto vehicle whilst one French and two Japanese craft contributed to a "Halley Armada" on its last passage around the sun. It returns every 76 years, and unlike 1986, let us hope for a better view from Earth during its next perihelion in 2061.

Elsewhere, Deep Impact fired a probe of its own onto the surface of Comet Tempel 1 (9P/Tempel) on Independence Day 2005, and in another mission, Stardust flew within 2,000 miles of asteroid 3355 Annefrank, later passing through the coma of comet Wild 2 (81P/Wild). A tiny sample of cometary dust gathered there was dispatched in a module home to Earth and as planned, this was recovered intact. Five years later Stardust paid another visit to Comet Temple 1.

More recently, the European Rosetta spacecraft landed its Philae probe on Comet 67P/ Churyumov-Gerasimenko, transmitting *in situ* data during its perihelion passage in August 2015. In origin the body is probably the result of a joining of two comets. The resulting rubber duck appearance drew droll comments but the lander Philae made less than a successful touchdown.

The Dawn spacecraft's paths around asteroid 4 Vesta provided valuable new data about the second biggest minor planet. Monitoring its approach over months was riveting as the pictures became clearer. Apart from fine images of surface features, measurements convey that it is indeed differentiated, meaning that it possesses an intact iron core, mantle and crust of sorts. This layering in structure was anticipated for the larger asteroids but in the formatively deep past any further growth by accretion was likely stunted by the gravitational influence of Jupiter.

By remarkable chance, little bits of Vesta have found their way to Earth. Many such tiny pieces of material, often cometary in origin travel in space as meteoroids, can enter the atmosphere as meteors and occasionally survive to the surface as meteorites (Greek *meteoros* for high in the sky). Apart from getting our terminology correct, the mechanisms of such orbital transfer are also better understood. Without trying, we earthlings have found ourselves with genuine lunar and Martian meteorites in our possession — they wandered in space for hundreds of millions of years before arriving here by chance. Now, there is cost effective space research!

The Dawn spacecraft gently powered out of Vesta's orbit in September 2012, transferring its solar orbit to meet with the largest asteroid 1 Ceres in Spring 2015. Dawn is ion-propelled; building velocity over time rather than big propulsive burns of chemical rockets in space. Big boosters are still necessary to get the vehicle off the ground but ion propulsion becomes a viable by taking time to accelerate as a trade-off for better fuel efficiency in flight. (Apparently, "ion drive" was first uttered in Star Trek.)

We are also adept at gravity-assisted acceleration, using Earth's and other planets' gravitational fields to gain some free speed by a slingshot along their trajectories. The taxpayers will go for that too. Dawn is one of a series of the NASA's cost conscious Discovery Program.

Much further afield, the ESA's new Gaia mission is intended to make a 3D catalog of over a billion astronomical objects. Imagery from the familiar

Hubble space telescope has been breathtaking in both depth and clarity. The Hubble deep field has broken the distance record several times over, the pictures often featured in the media. The deep field images show whole clusters of galaxies.

Returning to New Horizons, NASA's website had a moving countdown busily clicking away as we moved towards our first encounter with Pluto. Earlier, Hubble had located a small fourth and then a fifth small satellite. For an ongoing mission with New Horizons we hope for premier visits to one or possibly two Kuiper belt objects lying beyond Pluto's orbit. This has been refined to a prospective encounter with the KBO 2104 MU69 in early 2019 if all goes well.

The Kuiper belt is an elliptical (?) and thin circumstellar disc estimated to lie 30–50 AU from the sun, a range equal to 2.5–4.5 billion miles. The members of the closer asteroid belt mostly follow orbits between the paths of Mars and Jupiter, a far closer 2.2–3.2 AU from the sun with several distinct groups. Their more rocky and metallic compositions equal a total mass some 4% of the moon but the Kuiper belt is probably twenty times its width, containing a mass greater by a factor of two hundred.

Let us also clarify the modernized status of Pluto. By new definitions it is classified as a dwarf planet. The decision to relegate Pluto raised some ire in the astronomical community and elsewhere viz. the International Astronomical Union's Resolutions of August 2006. It had been regarded as the ninth individual world since its discovery in 1930.

As a dot among fields of stars it had twice gone unrecognized but these photographic plates were the work of Lowell Observatory where Pluto was finally found by painstaking search and comparison photos. It was a feat of individual telescopic skill by Clyde Tombaugh, despite the long sought Planet X proving disappointing in size. Pluto was decisively not another gas giant planet and we now place it as the prototype among a whole class of Kuiper belt objects rather than a ninth planet. Dwarf planets are defined as:

(1) Being in orbit about the sun.

(2) Having sufficient mass to assume the hydrostatic equilibrium of a nearly round shape.

(3) Having not cleared the neighborhood around their orbit.

(3) is the essential difference between a dwarf and classical planet, the latter having cleared their respective paths about the sun unlike the members of the described Kuiper belt. Methodically if controversially, this now includes Pluto and reduces us to eight classical planets. Uranus and Neptune were discovered in 1781 and 1845 and are not part of traditional planetary lore.

Such clearings do not include retinues of moons of the outer planets and it is not a matter of pure size. Jupiter's moon Ganymede and Saturn's attendant Titan is slightly larger in diameter than the planet Mercury. (Note that both are less massive than Mercury.)

Asteroid 1 Ceres (diameter 588 miles) is regarded as a dwarf planet and the other asteroids are still designated minor planets. The latter was decided in the 19th century when they were first found and stands in our methodology. But a case for the very largest and broadly spherically shaped asteroids receiving such status does exist. For now, the officially recognized dwarf planets are Pluto and Ceres, plus the KBOs Haumea, Makemake and Eris. Let us move up to the cutting edge of knowledge for the furthest known bodies of the solar system.

Scattered disc objects, Trans Neptunian objects or even inner Oort Cloud objects are other appropriate terms in this formative era of discovery that commenced in 1992, when a KBO beyond Pluto was first found. Their ranges, periods and inclinations far exceed those of the classical planets and it will be some time before any detailed imagery or full details become available.

Clearly, there is much to learn about their sizes, distribution and most importantly, their origins and we should be optimistic about future techniques. They are surely primeval, and some definitely possess gravitational resonance with the orbital period of Neptune, such Pluto's own 2:3 relationship with Neptune. They could be suitably dubbed "plutinos." Possibly, the components of the Kuiper belt were directed by the outward migration of Neptune long ago.

Likely, Neptune's largest satellite Triton is a large KBO captured into a retrograde orbit and Saturn's moon Phoebe originated in the Kuiper belt. The latter's low density, strange sponge like appearance and another retrograde path about its primary indicate this for this largest of Saturn's outer satellites. It was well imaged by Cassini but had been the first satellite discovered by photography in 1898. Alternatively, Phoebe could be a centaur, the class of body that has characteristics of both asteroid and comet and move with semi-major axes between those of the outer planets. (The semi-major axis is essentially the radius of an orbit at the two most distant points of that orbit.) Currently, there are an estimated 44,000 centaurs with diameters larger than 1 km. The sun has a huge and diverse family.

By pure estimation there may be 200 bodies in the Kuiper belt the size of Pluto and as many as 10,000 in total, some possibly as large as an inner planet.

90377 Sedna subsequently introduces "sednoid" for bodies with perihelia (the closest point to the sun) greater than 50 AU and a semi major axis exceeding 150 AU. Sedna itself is some 1,100 miles in diameter with a period of some 11,000 years. It lay three times the distance of Neptune when first found.

Such bodies range much further than the assumed boundaries of the Kuiper Belt and we currently know of 589 bodies with aphelia (the far point from the sun) greater than 103 AU. Two KBOs with retrograde orbits are known.

In late 2015 the Japanese Subaru Telescope at Mauna Kea discovered what is the most distant known object in orbit about the sun so far.

The yet unnamed V 774104 lies at 103 AU or 9.6 billion miles. These are reliable first estimates but observation arcs based on the earth's daily parallax cannot allow precise orbital data immediately. A reflectivity of 15% at 24th magnitude could indicate V 774104 to be 300 miles in diameter. Along with 2012 VP_{113} it comprises the third ascribed sednoid.

Possibly, the Kuiper belt produces some quantities of comets but the much further Oort cloud is posited as a far more productive source. As the scattered disc is probably dynamically active and the Kuiper belt relatively stable, the former is the more likely origin of at least short periodic comets that are usually at a shallow angle to the plane of the ecliptic and move in generally prograde directions.

Conversely, long period comets vary widely in orientations and about 50% of them pursue retrograde paths at widely varying orientations indicating an origin in the further Oort Cloud. As a minor time capsule I invite my readers to see what progress NASA's proposed Whipple mission has actively made from the dream sheet to cruising the actual Kuiper belt by now.

As for newly imaged Pluto, the IAU approved "Tombaugh Regio" as the name for its most prominent surface feature. Surely it is deserved. After 85 years of mystery it must have been ready for its close up. Fittingly, a small amount of Tombaugh's ashes are aboard New Horizons, just as some of Gene Shoemaker's made it to the moon.

Applying extreme optimism, the craft could conceivably assist in the dynamical problems still associated with the exact paths taken by Uranus and Neptune. In retrospect, Lowell's proposed Planet X needed seven times the mass of the earth to produce the effects and we wonder how he was able to estimate the orbit so well. Probably it was great luck and some sort of poetic redress for his errantries about Martian canals. Pluto only has a mass 0.002 that of the earth so the greater mystery still stands.

All such initial encounters are the lifeblood of planetary science. When the images of an unknown world's real face are first revealed to curious human eyes they convey a drama like no other. What of planets beyond the solar system?

The Kepler satellite mission is providing an enormous amount of new information on extra solar planets. There are strong indicators that they are common, indeed the norm about other stars. Hence, potential abodes of life exist in substantial numbers.

But pause, we must not get ahead of ourselves here. What we have established, at length, is the reliable existence of over 2740 extra solar planets and strongly counting as the data sort continues. The hunt is on for "superearths" as planets bearing some broad similarity to our planetary home. One of the most revealing patterns lies in so many "hot Jupiters" lying close to their parent stars.

At last there are hard data on exoplanets orbiting other stars after long speculation and a few false starts. The very first definitive extra solar planet was discovered orbiting a pulsar in 1992. Three years later a planet was deduced to orbit the star 51 Pegasi, identified by the radial velocity method of ground based observation. These are the gravitational effects that planets have on their parent star, a measured weave of a rotating dumbbell effect. The other telescopic technique includes plotting the light curves shown by the chance occultation of companion bodies. Fruitful as they can be, Kepler is a major advance on these methods.

Locating a planet circling the binary star system labeled Kepler 16b and a prospectively Earth-like planet orbiting the star Gliese 581 effectively processes science fiction into reality. The Kepler mission website is a trove of approximate planetary characteristics and orbital data, a strong work in progress.

The most peculiar light fluctuations of the star KIC 8462852 could be an alien megastructure. This is the most imaginative interpretation for Tabby's or Boyajian's star, perhaps the most mysterious in the Galaxy for its unique light curve. Putting aside an extraordinary behavior of the star itself or some intermediary agent across 1300 light years affecting only this star, the objects in its orbit cannot be easily explained. However, extraordinary claims require extraordinary evidence. The cause being a vast orbital disc of dense but fine particles of dust now looks the best hypothesis.

As for the Kepler mission, visit http://kepler.nasa.gov/Mission/discoveries periodically and see if you can keep up with it, earthlike worlds and all. The existence of a planet orbiting Proxima Centauri is further indication of their prolific existence of planetary worlds in other star systems.

For our latest achievement, the Juno mission should reveal more on the structure and huge atmosphere of Jupiter. This is one of the largest unmanned space vehicles yet built and after two years its path in space received gravity assist from Earth, sending it on its way. In July 2016 and after five years and two billion miles of flight it successfully entered a polar orbit lasting an initial 53 days. We have hardly scratched the surface or rather the cloud tops of Jupiter's prodigiously dense atmosphere although the Galileo mission made its grand finale with a deliberate dive into the great Jovian clouds as Juno is finally planned to do.

According to myth, the king of the gods drew clouds to hide his mischief but his wife Juno could pierce the veil and see Jupiter's true nature.

For sheer distance the most ambitious ongoing projects remain the two Voyagers. Launched a few weeks apart in 1977 they graduated to become the most far-flung feats of human engineering yet attained. They are cast forever into the cosmic ocean having long since served their prime objectives in flybys of the four outer planets and many of their moons. They are already further away than V 774104. The record will stand for any foreseeable future and until some highly futuristic fluency of interstellar voyaging is within our technological grasp. In a way they are the most important "first" in un-manned spaceflight.

Having surveyed Jupiter and Saturn over 1979–1981, Voyager 2 contin-ued to Uranus, reaching it the same week as the Challenger space shuttle disaster when enthusiasm was painfully muted for high jinx in space. Just over three years later the mission scored another spectacular success with its scheduled encounter with Neptune. For a long time to come the close up pictures of Uranus and Neptune you see in texts will came solely from Voyager 2.

Where are they now? Voyager I is currently approaching 13,000 million miles from Earth, definitively the distance record. After 35 years of voyaging it was announced that it entered interstellar space in September 2012.

In a different direction Voyager 2 presently lies some 11,000 million miles away and of course, both are still moving out. Earlier, Pioneer 10 and Pioneer 11 had been the first visitors to Jupiter and Saturn but we are no longer in contact with them, allowing only estimates of their positions. Despite a head start in launch dates, the Pioneers are actually less remote than the Voyagers with which we are still in touch.

The point is that these vehicles have sufficient speed to permanently es-cape the sun's gravity and by any working definition of a starship, five such craft have already been sent on their ways. However we define its boundaries they will ultimately leave the solar system entirely. Note that solar escape velocity is just over a million miles an hour from the close vicinity of the sun but considerably less further out. New Horizons was the first probe to be launched into a path of deliberate solar escape velocity and its expended booster makes for the sixth human built artefact to ultimately reach inter-stellar space.

Watching that Voyager distance data is a thrill in itself. The figures pile up by the second. Some onboard subsystems are being partly closed out but power and some useful data from four onboard instruments should last until 2025. It is now formally named the Voyager Interstellar Mission their speeds of approximately 36,000 mph as of late 2017 is sufficient to attain this great goal. A picture of a Voyager appears on U.S. passports.

This we *will* be remembered for and as a remote possibility, they could be the means by which some other intelligent creatures first hear from us should they ever find these space-borne artifacts of spacecraft engineering. Like a message in a bottle in the ocean of space this is a long shot at communication with an alien intelligence somewhere out there. Suitable greetings are inscribed on plaques mounted on the Pioneer craft and a more sophisticated sample of its sounds and music are attached to the Voyager vehicles. Recorded on a gold record are thoughtfully assembled murmurs of Earth like man's greatest hits.

Voyager 1 was crossing the termination shock lying approximately 94 AU from the sun in December 2004 and Voyager 2 at 84 AU in August 2007. The telemetry was still good enough to deduce this as they passed through the zone where the solar wind is slowed to subsonic speeds by collision with the local interstellar medium. Here the solar wind of charged particles from the sun is equalized by the incoming stellar winds. Beyond is the abyss of interstellar space.

We estimate that Voyager 1 it will pass within 1.7 light years of the star AC+793888 at a distance of 17.6 light years in 40,000 years' time. Voyager 2 is heading in the direction of Sirius but it is a matter of 296,000 years before it passes it at a distance of about 4.3 light years.

What is the very furthest extent of the solar system? The Oort cloud (or Öpik-Oort cloud) is thought to be a spherical cloud of icy plantesimals and material whose inner periphery commences at an estimated 5,000 AU from the sun ranging out to perhaps 100,000 AU. This is a thousand times the distance of the Kuiper belt. This proposed outer periphery equals about one third of a light year or one fourteenth of the distance to the nearest star. The Voyagers will reach the proximity in about 15,000 years.

We still speculate on its part in the formation of the solar system, apparently an outer residue of ice and dust left apart as a periphery to the development of planetary bodies much closer in. The model sounds good enough. Its scale is too great for material originating in the closer asteroid belt to have been driven so far out by the gravity of Jupiter as may be the case with the Kuiper belt.

The Oort cloud certainly provides something of the vast repository of cometary material before they are beckoned sunward by gravitational tugs, possibly nudged by stellar neighbors. Untold quantities of comets potentially result.

What are comets? These loosely bound amalgams of cold hydrocarbons, rock and ice initially take long paths of approach as they plummet closer to the stellar fire and gravitational center of the local planetary system.

Drawing closer they become heated and activate, casting off material over timeframes very brief in comparison to their orbital periods. Their nuclei are up to a few tens of miles in size, surrounded by an awakened gaseous coma during the plunge at top speed and closest approach. Heat causes their volatile components to sublimate into space resulting in impressive ion and dust tails.

After passing through perihelia they generally head back into planetary space, sometimes developing less eccentric orbits over multiple revolutions. Halley's Comet ventures beyond the orbit of Neptune, reaching aphelia of 35.1 AU in late 2023 when it begins to turn back. Tiny pieces of it are visible annually with the Orionid and Delta Aquarid meteor streams so one can see extracts form the parent body twice every year. It was unfortunate that Halley's perihelion was on the other side of the sun in 1986. Two centuries is the accepted demarcation for long/short period for comets.

Alternatively, some gain enough speed to escape the sun entirely, passing close but once as they plunge into the greater chasm. Others split up or are torn asunder by the stresses of close passage as did Comet ISON. In 1994 the collision with Jupiter's atmosphere by Comet Shoemaker-Levy 9 was another special show.

Being low mass they are short lived by cosmogonic standards, cosmogony being the branch of cosmology describing the emergence of the solar system. Meteoric material has its sources in both expended and surviving comets and known asteroids and the earth encounters ribbons of debris shed along their paths, often at regular points and hence at predictable dates throughout a year.

As "dirty snowballs" comets are the commonest objects in the solar system but in terms of density, they are proverbially the closest thing to nothing that can still be something.

What of asteroids? Again we shall pass on standard description to ask instead what real threats they represent apart from tedious movie plots. We shall return to the matter in chapter 4.

The Danger Of Asteroids

As for "target Earth" the latest results of the Wide-field Infrared Survey Explorer indicate that Near Earth Objects are less common than previously supposed. We assure ourselves that 90% are basically located although there is clearly a lower practical limit from any ground or satellite observation to find them all. This could prove tricky with nasty surprises lurking.

In the wake of the Chelyabinsk event of February 2013 the whole discussion of has been reopened. Unconnectedly, the small asteroid 2012 DA 14 passed within 17,200 miles of Earth the same week, closer than the orbits of some geosynchronous satellites.

With the Chelyabinsk meteor, the nature of its mass, speed and approach have been confidently gauged. Studies include its former path in space and the recovery of surviving fragments after its troublesome arrival. There is plenty of footage of its abrupt and blazing appearance as recorded by dash cams and security cameras but the joy of astronomy evaporated in the hail of flying glass, damaged buildings and hundreds of people being hurt. Observing meteors is normally far more sedate.

The statistics include an explosion at 16-19 miles altitude by a meteor entering at 40,000 mph with significant smoke trails. It was the arrival of shock waves of compressed air about three minutes later, not the strikes of fragmented meteors that did the damage. With an estimated mass at 12,000 metric tons releasing 20-30 times the energy of the Hiroshima explosion things could have been much worse.

One good candidate for the parent body of the Chelyabinsk superbolide is the Apollo type asteroid 2100 EO_{40} which was 40 days past perihelion at the time This is good news for the science of celestial mechanics and meteorite collectors but a relatively small/dense/high speed meteor could appear out of the sky at any time, surviving all the way to the ground to cause widespread damage by both direct impact and accompanying shock waves.

Spectacular meteor showers like the Leonids storm of 1833 gain mention in the annals but there are mercifully few accounts of impacts causing casualties. Yet for cosmic bodies abruptly striking the earth with the force of a nuclear bomb, no such explanation has ever been reached for the Tunguska event of 1908, the greatest natural explosion in recorded history. Studies suggest that it was neither an asteroid nor comet that so dramatically appeared early one morning in the wilds of Siberia yet left no material trace of itself. (See my own *The Tunguska Event or The Great Siberian Meteorite*.) That mystery endures for the ages.

Normally, anything of significant size rarely approaches closer than the moon but meanwhile, the occasional zip of meteors and occasional fireballs entering the thickening atmosphere are occasionally seen in a clear night sky. Regular showers are eagerly awaited in the observing calendar — that leisurely pursuit for backyard astronomers patiently deployed in deckchairs hoping for a display of natural celestial fireworks.

As for streams definitively traced to parent bodies in source, examples include the Perseid meteor stream descending from Comet 109/P Swift-Tuttle. It follows a 133 year orbit and the earth's path encounters trails left from it every July to August. It forms a reliable stream to watch for. The Geminids derive from asteroid 3200 Phaethon that has a very close perihelion of 0.14 AU. During its orbit of 1.43 years thermal fracturing regularly causes its 3-mile body to further crack and crumble and tiny strewn pieces cross our own orbital path in mid December. We question whether there is such a thing as a truly sporadic meteor.

Asteroid 2012 TC4, an estimated 56 feet in scale passed at a mere 59,000 miles from Earth in late 2012. Of course, such things are no more common than ever, we are merely much better at detecting NEOs nowadays and the list is impressive. We would be wise to monitor them as best we can, an example of something we cannot afford not to be doing. Obviously, early warning of a potential disaster hopefully would allow the time for prevention. 99952 Apophis might be a little too close for comfort on 2023 April 13. We have already assured ourselves there will be no Armageddon this time around although it will be naked eye visible. Asteroid 4179 Toutatis is another Apollo class "Earth grazer."

As for dealing with a real threat, nuclear warheads might actually be of use, launched to intercept and destroy anything on a collision course with Earth. There are some impressive studies of potential scenarios on file. What we have gleaned is that deflecting its path by space tug or gravity tractor would be safer, meaning going out to meet a killer asteroid and give it some sort of a nudge in advance of catastrophe.

This assumes there is time to divert the intruder, a major assumption in addition to making ready the equipment to pull this off. Blowing it to bits could compound the problem of ground impacts, showering us with separate pieces of the same menacing mass on the same paths.

Our atmosphere provides good but limited protection and we conclude how altering its trajectory in space would be a safer course of action. As for detection, we are equipped to adequate standards by now to see danger coming. We hope.

The forthcoming Double Asteroid Redirection Test by NASA's proposed Asteroid Impact and Deflection Assessment (AIDA) designed to rendezvous and directly engage 66803 Didymos is a good step forward as a dress rehearsal for what could be a dangerous threat one day.

A World Of Our Making

The effort to understand the universe is one of the very few things which lifts human life a little above the level of farce and gives it some of the grace of tragedy.

— Steven Weinburg. Epilogue of *The First Three Minutes*

As a species, we must seek a better comprehension of our own origins. In order to construct any relationship to the cosmos, where do we fit into the context of life on Earth? Specifically, when did we first become fully human?

Our deepest sense of identity is still up for detailed scientific grabs and we must sift the dust of ages to sketch out an authoritative picture. Of course, it is a matter of carefully assessing the available evidence like the better interpretation of those human-like bones and fossils to analyze our paths of our descent. It is time to look back.

Tracking some major DNA trails is an informative approach for the topmost layers, the spreading out of the human race from our assumed African origins. Those studies probe our true roots in at least the geographic sense. DNA stands for deoxyribonucleic acid, which is the molecule that encodes the genetic instructions used in the development and functioning of all known living organisms and some viruses.

Sharing so many genes with other apes never sat well with the anthropocentrically minded. Yet is an obsolete vision that man was directly placed in a whole and current form into the pride of place in nature. It still raises indignation among religious fundamentalists holding humans created entirely separate to all other life, pointedly excluding the large living apes as our biological relatives. Science occasionally shatters long held illusions. It banishes superstition and ignorance to call forth a better vision and I see no exceptions in the world we have made.

As always, we must follow where the evidence leads and construct theories based on empirical facts, findings and sensible deductions. In basic terms, the scientific method holds that theories must either:

(A) pass the test of time or

(B) be amended or entirely scrapped.

That man is related to the apes is in the first category. The abandonment of the concept of a luminiferous ether pervading all space to allow the propagation of light in vacuum exemplifies the latter. Despite a long assumed presence in physics no sort of ubiquitous elastic substance exists. Conversely, as Thomas Huxley once observed, the great tragedy of science is the slaying of a beautiful hypothesis by an ugly fact. This is why a viable theory must be falsifiable.

We need to get over it that we are closely related to apes, simply because we *are*. Apart from obvious similarities, the supporting evidence of genes and DNA are conclusive. Our closest living cousins prove to be chimpanzees and orangutans, and we need to reexamine the cultural baggage inherent in thinking otherwise.

It was probably a precariously small group of *Homo erectus* (meaning upright man) striking out from East Africa that literally took the first steps towards globally flung human families. We have hints of a comparatively recent DNA trial along the paths these tribal nomads took across the Levantine Corridor and Horn of Africa into Eurasia an estimated 1.8 million years ago. Other groups continued further into Asia and northwest to Europe.

When do we first get to grips with any self-recorded story? The tiny tip of the monolith of time where we enter the story commences about 5,500 years ago.

H.G. Wells described the future as an eternal dawn of possibility and opportunity, but he also pointed out that man writes history in his own blood. Whilst at odds, both views seem momentously true, the future being a race between education and chaos. What a sad sense of contrast between the ways things are compared to how they could be for our species. If only our science matched our wisdom. It is perverse that we, the first generation to methodically think about the future, may not have one.

Science itself has proceeded in fits and starts with periods of serious stagnation and knowledge actually lost. It is only in the last two centuries that science and the applications of technology have been so very influential to the human condition. It was only in 1834 that William Whewell coined the word "scientist" to supplant such terms as "cultivators of science."

The rôle of the discoverer has not been consistently respected or even allowed in some times and places. Discoveries unwelcome to the political and religious status quo have occasionally rendered thinkers too smart for their own immediate good. Science and its applications began to play an increasingly important part and in astronomy we owe much to Copernicus, Galileo and Kepler — leading to Newton and Einstein.

More practically, applied technology brought permanent changes to our ways of life, defining this "developed" world of ours upward from the industrial and scientific revolutions. Due to our sheer weight of numbers, there is no way back to simpler times. Notably, only select portions of the globe enjoy its full benefits or have access to the positive achievements in the world of our making. Too much of it is still torn by strife, conflict and poverty.

Such transformations commenced humbly enough with the application of fire and the making of simple tools as the first stirrings of rudimentary technology. Peking Man, a type of *Homo erectus* employed such in caves 700,000 years ago and evidence from South Africa's Wonderwerk Cave reveals an attended hearth from over a million years ago. Such things distinguish us from other creatures. Implements for hunting and the use of fire may have directly saved us from icy doom when the great glaciers were abroad. Obviously, we survived the last big ice age. They characterize the long Stone

Ages before the rise of large organized societies and civilization, and meanwhile, our hunter-gatherer instincts have not changed at all.

Many energetic and colorful human personalities have emerged in the quest for knowledge. They have richly added to its repository with dedicated work and touches of brilliance. Great thinkers wield the broadsword of reason, displaying the courage to develop the very methods of science. The spirit of rational enquiry blazes across the ages, as does our deep drive to explore.

Progress has been badly dampened down at times to the extent of suppression, lapsing us back into stagnant states of existence. These detractions still occur in the 21st century. Sometimes they seem to be getting worse. Barbarism and superstition lurk but a backward step away.

Prior to the first scientific revolution, all scholars held the received Aristotelian and Ptolemaic systems above reproach. In the long intellectual slumber of Europe between the fall of the Roman Empire and the Renaissance, such interpretations of their works from the classical past accompanied by the holy word of the Bible were considered infallible sources of truth. In modern parlance, you'd better believe it.

Consider the following facts: Contrary to what was thought, the world is not at rest at the center of all things; and when bodies of different weights are allowed to fall, they do not move at different rates. This is actually easy to show.

Overturning the first assertion was integral to the "new astronomy" of the Renaissance, which concluded that the earth is spherical and moves around the sun. By simple experimentation, gravity is shown to act uniformly on falling bodies. By the 17th century, the old, established physical worldviews were proving seriously erroneous. In support of the sun-centered scheme of Copernicus, it was Galileo who provided the first observational evidence. He was first to apply a telescope to astronomical bodies and publicize his findings.

In the decade 1609–1619, Kepler formulated his three laws of planetary motion, his principal sources deriving from the vast body of practical observations bequeathed by his collaborator Tycho Brahe. (He was however the junior partner.) Newton's *Philosophiae Naturalis Principia Mathematica* of 1687 expounded among much else that the action of gravity is a combination of the masses of both an object and the earth. It certainly applies to the moon's motions just like falling apples.

He expressed this and much else in mathematical precision in his epochal works. The concept of gravitation being a localized curving of time and space was not discerned until the early 20th century. It is anything but obvious.

Whilst gravity can be calculated as a force, it proves to be so much more. Exactly how a force can act at a distance was long a mystery and Newton himself acknowledged that even its most precise description is not explanation. The most calculated *how* concerning gravity does not immediately indicate the *why*. *Hypotheses non fingo*, as he put it in the essay making up the General Scholium to the second edition of *Principia* in 1713. Most importantly for the new physics of motion, Newtonianism shows that the same laws apply to celestial and terrestrial bodies. The world is not ruled differently to the heavens around it.

Despite their intellectual honesty and the five centuries separating Aristotle and Ptolemy, their unassailable reputations crumbled badly in the light of freer investigations. They probably would not have approved how their systems became so institutionalized and dogmatized as to hamper new drives for knowledge.

For examples of this consider that the planets go round the sun, not the earth, as Ptolemy had expounded at length in the 2nd century AD. Similarly, Aristotle (384–322 BCE) asserted how objects move to their natural place, which is the central Earth. In a nutshell, the earth is not too big to move and it was another erroneous current in Hellenic thought that there is a radical difference between the materials making up heavenly and earthbound bodies.

Fate has dealt differently with enterprising thinkers and theorists over time. Societies have bestowed honors and recognition on some protagonists of new knowledge, greatly profiting and advancing from their achievements. Elsewhere they have been ridiculed, ignored, put on trial and occasionally killed for their efforts. In the harsh rulings of the Holy Inquisition and French Revolution respectively, Bruno and Lavoisier were judged heretical and reactionary and paid with their very lives. In a strange twist, Curie fell victim to the very radioactivity she was studying.

Progress has been anything but smooth but the expansion of empirical knowledge and its useful applications forms a real progression throughout our often jaded past. I find the story of science to make more uplifting reading than most other specialized forms of historical study. It makes meaningful sense to see what has been found out, who made it happen and how it all unfolded. One gets to meet some imaginative and curious people amidst whole new spheres of endeavor and accidental discoveries abound in the story. We have produced some astonishingly creative and investigative minds, some proverbially born before their time and with more than a little genius to apply. Similar credit should be accorded the writers and artists that made expression a stimulus to thought.

We now turn to the larger story of life on Earth that, in a qualified sense, culminates with humans as its most self-aware and accomplished product. I advocate the dignity of educated humility over the overblown sense of self-importance that held us back for so long. Thank goodness for the Renaissance and the scientific revolution for what they produced in the arenas of rational thought and knowledge.

Accepting that the progress of astronomy reveals the overarching insignificance of man, we can actively enjoy rationalizing the scales of cosmic time and space as a slice of empirical truth. This we will be undertaking in our discussion.

There is an imposing adjunct to the humble place we occupy in the universe. Apart from the hypergiant stars, expanding galaxies *etc.* and their prodigious energies lasting billions of years there emerges a most notable fact. For all the matter and energy abroad, the most complex state of matter known to us is the human brain, and it bears enormously greater potential of function than we ever consciously use. Some would give their heads the very crown of creation. Traditional belief systems have long asserted a very special status in creation for us. So what is man's place in the universe, given what we now know of its staggering scale?

Bertrand Russell once noted that any concept holding that man is the center of all things is a dangerous illusion best cured by a dose of astronomy. He also looked diligently for evidence in favor of the statement that man is a rational animal, only to find evidence to the contrary as the world plunges continually further into chaos.

Let us take a look at this great natural world. Firstly, the biological mechanisms of natural selection promote adaptation and change, firmly ruling against the continuance of some creatures. We debate its causes but on clear evidence the death sentence has been wielded more often than not on whole species in Earth history, there being far more extinct than extant creatures overall. Few living creatures have close ancestors going back for more than 500 million years.

Even applied evaluations of the fossil evidence provide much less than a coherent picture of the complete past, including recent upstarts like man. It is a complex jigsaw and we have but few of the pieces. It is clear that changes are going on over great timescales. Given the scientific breakthroughs of the past few hundred years, we may one day understand better how to account for the remarkable biodiversity we see.

The origin of life on Earth is mostly still a mystery. Further afield, some suggest that it might have arisen on Mars only to die out. What a turn of events we could anticipate, if it was proven that there had been life on Mars long ago that did not survive to the present. It is possible that it did not survive because the environment turned hostile. Possibly, Mars might have lost its greater atmosphere and reserves of liquid water, slipping into a dry coldness no longer supportive of life. Indisputable fossils or biochemical traces, found perhaps very deep in the Martian crust, would be convincing evidence and this is a viable investigation. It could be the foreseeable next giant leap.

Let us not overreach. It is established that man has not been around long in the great vista of Earth history, a conclusion we shall examine in detail. What can we reliably say?

Firstly, we must accept that great changes are inherent in the great natural chain of events we are part of; and secondly, there can be no guarantee of continued success, especially being based on a single planet. Here lies one, if not *the*, prime long-term purpose of space exploration. The full-on colonization of at least near space seems an imperative of the proper development of a technological advanced species. Having all your eggs in one planetary basket sets a dangerous limit on continued existence should disaster strike the lone planet of residence.

The Path Taken

In the spectrum of human activities, science is richly cumulative. Let us accept that intelligence begets intelligence and that knowledge is power. Scientific research grew from risky individualism to something of a gentlemen's hobby club in post-Renaissance Europe, developing up to our era as government operations and full commercial enterprises. We shall examine some of the nuts and bolts of the received human condition and the cause and effect of our most recent alleged progress.

The most concerted technological efforts of the 20th century were the Manhattan and Apollo projects. These were the most focused and ambitious scientific endeavors ever undertaken. Some of the brightest, best-educated minds working together and highly motivated, funded by major government investment, they achieved two contrasting results, respectively: the debatable "success" of creating the atomic bomb, and separately, to win the space race of the 1960s.

In the West we owe much to our later classical Greek and Roman forefathers. For practical example, modern English language comprises of 10% Greek (Gk) and 40% French or Gallic Latin (FrL) as dictionaries give — indicating that the very words we use are living descendants of their past triumphs and disasters. Half of our very words were bequeathed to us from them.

The very term "history" is late Middle English in origin, deriving from the Romans' Latin language via the Greek *historikos* meaning inquiry. In immediate times, the general level of science and history education could be improved, as surveys embarrassingly show concerning the non-comprehension of so many vital things that have shaped us.

We tend to take science-enabled life styles for granted nowadays, yet it is only in recent generations that information technology, high-speed transportation and instantaneous communication came about. Now there are serious questions about how to keep it all going, because the earth's resources are strictly finite. We fear for our global carrying capacity and the ruinous impact on the environment. Of course, telling people to settle for less and consume only minimally is not good for business or attracting votes. We leave our children many challenges.

Empire building based on wealth, power and territory was nothing new, but the material means became far more assertive after the Industrial Revolution. Land grabbing abroad was particularly popular among the powers of Europe, first enacted as early as the late 15th century in the pursuit of trade routes and exploration.

It was to create obligations for the future and forms of blowback absolutely not anticipated (not that the perpetrators cared one jot at the time). They saw nothing but profit in the short term and domination by their political masters for the longer outcomes. The voyages of Columbus, Vasco Da Gama and others were hardly in the spirit of science and peaceful development. They called the Americas the "New World," as if there were no indigenous occupants. The crimes of our forefathers include the forced creation of an African Diaspora and the bitter fruits of slavery. Here, the statistic of twelve million people methodically uprooted over four hundred years may be an underestimate. As a result we face a still-growing social burden to this day.

Philosophy was reignited by our dear René Descartes (1596–1650), who presaged the European Enlightenment, in academic thought at least. No longer was it heresy to propose that nature follows laws as opposed to bearing witness to God's glory, *et al.*, and implicitly that "no further details of enquiry are necessary or permitted."

Descartes did have to take care lest he collide with religious authority; even so, he still ran into occasional trouble. Galileo encountered more severe opposition by the Church, which eventually silenced him. Kepler had mostly the good fortune to flourish in Protestant Europe, although he met censure at times. Offending established authority slowly became less dangerous over time.

The concept of scientific determinism was most fully strengthened with Newton's breakthroughs. He is a very special character; his monumental *Principia* giving rise to a whole paradigm of methodical and exacting thought and calculation. With him, mathematics and motion came of age and classical physics was established. Such were his achievements that there was criticism for leaving little for the Almighty to do and science books still make more references to him than anyone else. Note that Newton was a biblically based theist of hostile independence and that despite his unique scientific contributions he took other arcane knowledge to the grave. We will never get to the bottom of it.

For the mass human condition, the Industrial Revolutions were to cause huge and irreversible social changes. It too commenced in the West, specifically in Britain with steam power and the machinery of production leading the charge. Coalfields provided the very fuel of this rapid development. In the unrelenting pursuit of product and profit, agricultural-based serfdom was updated into wage slavery. In the exchange of farms for factories, a few commentators expressed at the time that it was not for the better.

Yet it was a real break with the plodding and unchanging past, especially in mechanized efficient productivity and growing political clout. It released forces in expansionism to crops, livestock or goods exchanged in simple patterns of trade. Investment replaced barter and the forces of capital wholly superseded their exchange.

The whole idea of a national economy or GNP is relatively modern, as are population censuses for a nation. The new science of economics academically commenced with Adam Smith's *An Enquiry into the Nature and Causes of the Wealth of Nations* of 1776, the same year that the American colonies declared their independence.

Accompanying this energetic political and economic expansion, the clock hands of human history seem to speed up. Railways and factories allowed hugely bigger and profitable production and superior means of transportation for both goods and people. Population swelled in a relatively short time for the newly industrializing societies, cities growing prodigiously. Now there was increasingly mechanized and efficient output plus far greater specialization in daily tasks and the accompanying paid employment. Farming and cottage industries as traditional forms were partly left behind. Mechanical devices beyond mere hand tools and labor form the foundation of industrialized society. Production was increasingly prompted along the bold lines of supply and demand and naturally, the motive of profit. Like usury, these were concepts absent and even disapproved of in earlier times. It was a major break with those traditions that sent us on our industrialized way during the expansive European 19th century. Skilled labor and cash flow became essentials.

Technologically, we acquired godlike abilities compared to the past, and we continue the upward trend. Huge changes in the human condition have ensued. Empires both nationalistic and commercial were hammered out by what has been termed steel and will, capitalism and a Protestant work ethic. Let us endorse "first world mentality" to supersede Max Weber's phrase here, having engineered whole industrial revolutions. It is not merely how people produce things but how they now think about the world, pointedly from the standpoint of their religious affiliations. (See *The Protestant Ethic And The Spirit Of Capitalism* 1905. It was a real eye opener at the time.)

In the West, the Victorian English felt that destiny had led up to them, that they were thus far its finest product. Contemporaneously we had the expanding American republic and, in the later 19th century, a unified, industrialized Germany. It was the ascendancy of the West.

However, this was like opening Pandora's box. Twice in the 20th century we plunged into global wars, killing millions and threatening to destroy whole societies. Improved firearms, tanks, aircraft, artillery and poison gas had entirely replaced cannons and charging cavalry on land. In 1914 warfare took to the skies whilst at sea, steel battleships; submarines and torpedoes prized naval warfare away from wooden ships and cannons. How could the European nations, long the leaders in culture and material progress, reach such an impasse? Today, again, we see that the immense progress of recent decades has brought vastly improved living conditions to millions, but mankind has been at least as creative in developing technologies of mass murder and in inventing excuses to use them. In this sense, it is difficult to say whether we are doing "better" than ever.

We will not dwell on our shortcomings, horrifying as they sometimes are. Let us instead find better uses of our energies and set the stage for a comprehensible look at the cosmos. This could be described as all there is, was or ever will be with emphasis on the cosmic sciences to guide us. As Sagan described our pale blue dot of a planet, it contains everyone and everything we've ever know. We have several real images of the distant Earth in this revealing guise. A departing Voyager departing from Neptune's environs even managed a family portrait of the planets of the solar system as mere point sources of light in the void.

I prefer the dignity of humility to being overwhelmed by our patent insignificance. Call it the search for meaning through cosmology or, to borrow a title from Penrose, the road to reality.

I gleaned the term "metagalaxy" from an earlier edition of *Norton's Star Atlas*, but the term gets little use and has now been eliminated from the very source to label the whole universe. "Cosmos" seems preferred nowadays, first used by Pythagoras as meaning "starry firmament." It came into Middle English from the Greek word meaning the good order of the world and entered use in the translations of Von Humboldt's treatises in the mid 19th century.

> The Greeks have borrowed a name for the universe from ornament on account of the elements and beauty of the stars. For it is among them kosmos...for which the eyes of the flesh see nothing fairer than the universe.

— Isadore of Seville (d. AD 636)

Chapter 2. The Sands Of Time, The Colossus Of Distance

The Big Time

We all know what time is until challenged to give a definition It is most simply described as the measure of duration and the temporal flow that organizes a past, present and future. It prevents everything from happening at once.

Time's biggest impact on the human experience is its irreversible passage because it claims us all in the end. Every person will be consigned to the past with regrets sometimes outweighing their dreams. Our lives are lived forwards but learned backwards and whilst we cannot live in any moment but the present there never seems enough time.

A mere century sees a virtual turnover of the world's living population, sweeping most of us into obscurity. Few people affect the course of history or carve actions into its edifice. Dwell a moment on the tyrants we could have done better without then more benevolent individuals leaving us too soon.

There are many layers and levels of time and it is the same with distance. Just as yards are impractical for navigation between continents, terrestrial miles or oceanic leagues are unwieldy to gauge the chasms to the least distant stars.

As the unseen dimension time complements the more tangible ones. Those recognized by the International System of Units as one of the seven fundamental physical quantities. We will shortly take a glance at its far-reaching implications. Hold onto your relativistic hats whilst we explore the deeper meanings of the depths of time in relation to space and physical matter.

We will commence by inviting select poets rather than men and women of science for instruction. E.T. Bell warned how time will make fools of us all in the end and Shaw lamented that youth is wasted on the young. Chaucer observes that time and tide wait for no man; and in Shakespeare's Macbeth, Banquo asks who can look into the seeds of time and say which grain will grow and which will not?

Elsewhere, a poem by Henry van Dyke's concludes that for those who love, it is eternal. This is a flight of romantic optimism because nothing can extend the strictly limited amount we personally have to spend. We all return to dust. Benjamin Franklin said that death and taxes are the only certain things, but at least, as Ashleigh Brilliant reminds us, living on Earth can be expensive but includes a free annual trip around the sun.

Entirely taken for granted, time was interpreted as the steadfast background to all things terrestrial and celestial. Our normal mental grasp encompasses no variations in its steady flow. Throughout traditional thinking and the reign of the Newtonian paradigm its passage was perfectly stable. Up to the emergence of relativity and quantum physics, past, present and future remained a stubbornly persistent illusion.

How do we perceive practical time? Sunrises and sunsets, the motion and phases of the moon, the cycle of the seasons — all provide regular beats in a system of high temporal order and hallowed repetition. The sun and moon seem great natural clocks in themselves with many observances and rituals attaching to them. We all have a certain body clock and a daily circadian rhythm.

Let us get to grips with the practical units, those presented by the natural world such as the diurnal cycle. The division into 24 hours practically serves for the more accurate 23h 56m 4.1s required for a daily rotation of the earth. We travel around the sun in 365¼ days in the meaning of a year with the cycle apparent at any latitude or climatic zone. Spring leads to summer and a chill accompanies the autumn air hinting of the winter to come.

Along its path the moon progresses from thin waxing crescent to full phase then down to a sliver of waning crescent in its orbit about the earth. The word *month* descends from moon. Its body is always 50% illuminated in space, but the quantity of sunlight on its face alters as the line between light and darkness — called the terminator — advances to full then recedes again.

A lunation is the complete cycle of phases and this average synodic period from one new moon to another relative to the sun averages 29d 12h 44m 3s. We utilize days, months and years based on the earth's rotation and the movements of the moon and sun. None of them make for tidy arithmetic units fitting each other. There is no reason why they should.

The causes of the seasons are straightforward to expound. The tilt of the poles of rotation lies at 23½° to the perpendicular of the plane of the earth's motion termed the ecliptic (Greek *ekleiptikós*). This inclination of 23½° equally applies for that of the equator to the earth's plane of movement. As a result, ever-shifting lengths of sunlight and darkness fall on its twin hemispheres over the course of a year with an unfaltering 12 hours of each at the equator. All other regions undergo a lengthening and shortening of days relative to nights generally accompanied by seasonal weather patterns and changing temperatures.

Towards the higher latitudes the ratio of daytime to darkness varies by greater amounts, maximized at the most northerly and southerly zones. The Land of the Midnight Sun is an endearing description for the very longest days (and nights) occurring within the Arctic circles. There is of course an equivalent Antarctic Circle placed 23½° from the southern pole in latitude.

The sun rises, crosses and sets daily but also travels a great annual path through the celestial sphere at a rate of just under 1° eastward a day as observed from the earth's great orbit. Its path is termed the ecliptic in the other sense of the term. Within the polar circles the circumpolar stars wheel around the celestial pole with no rising or setting and at the equator there is an entirely seasonal display. Most of the surface of the world lies between these extremes and sees something of both. The altitude of the north pole star in degrees is equivalent to a given latitude in that hemisphere.

The celestial equator is merely the line of the equator projected into the sky and the sun crosses it heading northward about 21 March as the Spring or Vernal equinox, still called the First Point of Aries where the ecliptic intersects the celestial equator.

One quarter of an orbit and three months later the sun reaches its far northerly position with the Summer solstice of 21 June. Following the autumn equinox of equal day and night around 21 September it continues south of the celestial equator to reach the Winter solstice at 21 December. These solstice points furthest north/south provides the longest/shortest days in the northern/southern hemispheres respectively. Comprehensively, the two tropics lie at 23½º north and south latitudes of the equator.

One endeavor of ancient civilizations lay in devising a serviceable calendar, often for the practical purposes of sowing and reaping at the right times.

In Egypt the annual inundation of the River Nile and the simultaneous appearance of Sirius in the dawn sky were greeted with much fervor. Understandably, this benevolent flooding of the Nile, accompanied by the heliacal rising of the brightest star assumed religious significance. The events were decreed as part of the grand order for the land of the Pharaohs and the heavens above. With a little imagination one can still see an Egyptian sky god in the guise of the constellation Canis Major.

The Babylonians were first to make meticulous observations, perpetrating the recorded dawn of astronomy. They also produced a more refined calendar than the Sumerians, primarily for those agricultural purposes. In full geographical terms they imagined the world as a disc afloat on a vast watery ocean and such impressions are conveyed in the surviving impressions of maps imaginatively produced by Anaximander and Hecataeus of Miletus.

In addition to the sun and moon, five other conspicuous bodies move independently of the fixed stars and keep to a narrow continuous band within it. These planets visible to the naked eye also change in apparent brightness with occasional loops, retrogressions and stationary points along their paths.

Like the sun and moon these wanderers move entirely within the belt of the Zodiac. It is actually an impression of the greater plane of the solar system as a projection of our own endless path. Observatory Earth is a participant, not an aloof or stationary observer of the planets' individual motions. What is seen is the combination of all orbits. Etymologically, the Greek words *astron* means star and *planetes* comes from wanderer. Zodiac derives from *"circle of animals."*

Maintaining an accurate calendar was a perennial challenge and the best attempts revealed flaws over passing centuries. This necessitated adjustments to be made, such as the adoption of the Gregorian calendar in the later 16th century.

Since the early Christian era, the precession of the equinoxes had inexorably shifted the precise position and date that the sun crosses the celestial equator. This imperceptible motion of the earth is like a slowly spinning top, primarily caused by a long-term effect of our mutual gravity with the moon and sun with a minor component from the planets. One result is that the First Point of Aries had since been carried into Pisces.

Hipparchus (190–120 BCE) first discovered the obliquity of the ecliptic, the inclination of the earth's equator in a truly remarkable feat of observation. (Tycho deserves a similar accolade for measurements that successfully pushed the limits of the unaided eye and it is unfortunate perhaps that he passed on so shortly before the invention of the telescope.) Some prior awareness of the changing positioning of stars must also have existed among the temple and pyramid builders of ancient Egypt.

None of this knowledge was available in the European 16th century but one earthly outcome, so vital in Christendom, was the date for Easter. It had been ordained at the Council of Nicaea, AD 325, to be the Sunday nearest the full moon on or after the Vernal equinox. It is formatively linked to the Jewish Passover festival but by AD 1500 the Spring equinox had drifted to 11 March and that far from the 25 March held in the existing Julian calendar. Things had to be put right.

In the revised Gregorian calendar a papal bull from that source of infallibility had the old Julian date of 4 October 1582 immediately followed by 15 October. One and a half millennia had passed since Caesar's edict reformed the existing Roman system as the year 709 AUC (*anno urbis conditae*) in what was retrospectively 1 January 45 BCE for the introduction of the Julian calendar. AUC refers to the semi-legendary founding of Rome equivalent to 753 BCE. In Europe, Anno Domini (AD) was not adopted until the 6th century.

We now know that the obliquity of the ecliptic is currently decreasing by 50.3 arcseconds over a year. This is equal to $1°$ over 71.6 years and a full precession requires about 25,700 years. Another effect is that the Pole star is not permanently placed close to the true north celestial pole.

Its current northern declination is 89º 19′ 8″ and, taking further account of nutation and aberration, it will be closest in early AD 2102 before the drift by precession continues celestial north towards Cepheus. Stellar proper motion is another factor for any star passing close to the wandering celestial pole over time.

This is why equinoxes and solstices shift and how there have been and will be other pole stars. Sometimes the job is vacant; and there is no equivalent South Pole Star. Polaris is about 433 light years away and its current positioning is quite fortuitous. It is a boon for navigation but in his day Pytheas of Massalia described the celestial pole as devoid of stars.

Time had always been perceived as some immutable agent, it's ticking so relentless that it allows measurements. What are clocks and calendars for?

This inflexibility was to undergo a serious overhaul. We will now graduate from poetic musings, Roman administrators and pious clerics to the findings of relativistic physics, commencing in the late 19th century. Here comes an introduction to the full relationship of time, space, matter and energy.

In the mundane world time holds entirely stable — with some vague ideas from documentaries that it passes slower if you go really fast or fall into a black hole. This is the case but neither hyper speeds nor monster gravitational fields intrude on everyday life. (Concerning our perception of time we will not examine mysticism, the domains of alternate states of consciousness or cases of people allegedly disappearing for decades.)

But we will assert that relativity is not beyond comprehension even if, as Einstein advised, it takes serious mental effort. In 1919 he put it thus:

> The special theory, on which the general theory rests, applies to all physical phenomena with the exception of gravitation. The general theory provides the law of nature of gravitation and its relations to the other forces of nature.

In the simplest terms these subsequently prove to be electromagnetism and the strong and weak forces of the atom. One is occasionally asked what this relativity thing is all about and here is an attempt on a daunting précis for the most salient features of both (A) Special and (B) General Relativity:

> A. Time slows down for the observer at hyper speed and within major gravitational fields. Ultra high speed also reduces the volume and increases the mass of the moving object. This

does occur in familiar and terrestrial motion but only at entirely negligible rates.

The speed of light (c) is a fundamental limit built into the very fabric of
the cosmos and here is the airtight case to prove it:

> B. For a body with rest mass to be hypothetically accelerated
> faster than light would not only require more energy than
> is extant in the universe but the body would become more
> massive than the cosmos itself. It would also occupy a vol-
> ume less than zero. In hypothetically exceeding the speed of
> light the slowing of time would pass beyond the stationary
> and go into reverse.

These conditions present unacceptable paradoxes and no such deals can
ever be struck for the perpetration of superluminal speeds although observationally, there are some intriguing illusions.

The incomparable Isaac Newton (1642–1727) passed over any matter of
variances in the flow of time. The first Scholium in *Principia* states in so many
words that absolute, true and mathematical time, of itself and from its own
nature flows equably without regard to anything external and that absolute
space similarly remains always fixed and immovable...

This now seems a simplistic view, asserting that there is nothing to *be*
expounded or analyzed. Time and space simply is, monolithically and un-
yieldingly. Scientific thought was held in this thrall for nearly three centuries
and such strict determinism characterized the "clockwork universe."

He also expressed that "this most beautiful system of the sun, planets
and comets could only proceed from the counsel and dominion of an intelligent and powerful Being" and he believed in occasional divine intervention
for the adjustment of some matters. This is one case for theism as the "argument by design" in the form that there can be no design without a designer
but is now outmoded that there is any "god of the gaps" *i.e.*, invoking divine
action here and there where full scientific explanation falters. We have long
since given up the practice.

Accordingly, time remained the all-encompassing and level backdrop in
a fully deterministic universe. It seemed cast in scientific stone, its unquestioned acceptance an unseen brake on deeper knowledge. It was an absolute
frame of reference although there were a few hints concerning some strange

effects before Lorentz, Einstein, Planck, Heisenberg *et al* enter the picture. Minkowski explored the implications of Special Relativity to the extent of that the fusion of space and time into spacetime is not uniquely Einstein's baby.

HG Wells was first to view time as a whole dimension but that space and time are one *and* that the universe originated in a single particle is first found amidst Edgar Allen Poe's *Eureka* (1849). Again, credit must be given when due but I definitely prefer to seek a Platonic spectacle of truth to macabre revelations.

Neither is the speed of light present in Newtonian equations although in *Optiks* he did pose whether gravity could influence starlight. This is an extreme precursor to gravitational lensing, the intriguing multiple and distorted images of distant galactic sources that are apparent over millions of light years due to the gravitational fields of others acting in the intervening distances.

Einstein's legacy is both outrageous and ongoing. Just recently, some detailed predictions of relativity were newly affirmed in studies of a peculiar pulsar and white dwarf companion star system.

Before we take on the origin of the universe, Sky and Telescope magazine once held a friendly brainstorming competition seeking a better name than "Big Bang." Terms like the Original Quantum Event are perhaps too technical to convey the Cosmological First Moment or Cosmogenesis. The Primordial Flaring Forth could be adapted but we seek something catchy and accurate. The term originated in opposition to the very concept and it took a while to catch on with its own supporters; but we seem semantically stuck with "Big Bang."

The Meaning Of Time

Let us glance at how the scales of time in the natural world were first properly revealed. Starting literally on the ground, the rock formations of the landscape and the processes at work there were being freshly appraised as early as the 17th century. When one looks at the posthumous impact of Copernicus and the careers of Galileo and Kepler in relation to the Reformation and Renaissance this is a massive understatement for the arts of fresh learning now abroad.

In mediaeval Europe the Mother Church strictly enforced orthodoxy in all things and new studies of nature were neither necessary nor approved. God's great handiwork was to be admired rather than investigated, a matter of praise but not methodical study. One cardinal admonished Galileo to better seek going to heaven than how the heavens go.

Besides heliocentrism replacing geocentrism aloft, it became increasingly clear that the actions and forces making up the face of the world are enacted over far bigger vistas of time than existed in the biblically based worldview.

The scale of the earth's biography needed drastic revision, requiring much more than some arbitrary 5,000 years of existence. Bede, Kepler and even Newton broadly corroborated this minimalist timeframe, which was the product of careful studies of the holy book with a few references to reliable secular dates in the ancient world. Note that the Hebrew year is currently Anno Mundi 5777. By purely internal reference they obviously have all their "begats" and such straight.

In 1650, Bishop Ussher arrived at a most erudite deduction, concluding that 23 October 4004 BC had been the date of creation. A brief summary of his work was promoted in a frontispiece to the King James Bible, a unique accolade from the Church of England for a scholar of any age.

He meant well and labored honestly, but a paltry 5,000 years does not encompass even recorded civilization, much less all that came before. When geology was still just a fledgling science, catastrophism (including Noah's universal Flood as a mode to shape the face of the world) began to lose its own ground. Beware, creationists.

At the opposite end of the scale, the invention of the microscope first revealed the realms of the miniature, including a host of living organisms too tiny for normal sight. The much-maligned Hooke first coined the term "cell" in *Micrographia*, the first book on the subject, published in 1665. Later, geophysics, astronomy and later nascent atomic theory required far greater graduations in the measures of space and time, both large and small. Let us sieve the sands of time to grasp the diverse scales associated. This too is worth the brain cells expended.

A little quantification is necessary. In calibrating time upwards, a billion or a thousand million years is termed a Gyr or giga year, most commonly Ga for giga annum, arithmetically expressed as 10^9 equivalent to 1,000,000,000 years. This 10^9 is simply 10 multiplied by itself 9 times in the numerical short-hand of power or scientific notation. One reads "billenium" in Soval's poetic style, and 10^{12} or a trillion is occasionally heard.

A million years is expressed in the same notation as 10^6 or 1,000,000, deriving Myr and mega-annum as Ma. A Ka or kilo annum stands for 10^3 as the passage of 1,000 years but outside geology, a "millennium" sounds nicer. For a statistic, there are approximately 7.29 x 10^9 people alive today. In power notation, a single integer and a stream of decimals designed for maximum accuracy precedes the powers. 7.29 x 10^9 = 7,290,000,000, allowing precision in big quantities.

To the other extreme of temporal smallness we progress downward far lower than the tenths or hundredths of a second applied to sports events or traffic flow in Seattle. They are not merely abstractions used by particle physicists and cosmologists although in a positivist approach, any quantified theory is merely a mathematical model to describe observations.

The meaning of the negative index such as 10^{-9} is simply 1 divided by the given quantity. The 1 divided by some quantity is termed its reciprocal so that 10^{-9} = $1/10^9$. It expresses very small quantities with the same convenience as the large. Adding or subtracting the indices further simplifies their multiplication and division. Consider the nano second at 1 x 10^{-9} or a billionth of a second. Relativistic effects genuinely emerge in telecommunications, pulsed lasers, atomic clocks and GPS.

The metric system lists tinier fractions of a second downward in brevity from the nano second to the extent of fifteen smaller magnitudes or factors of ten, all with catchy names. In the province of applied physics we read that a "jiffy" lies between the designated zepo and yocto seconds at about 3 x 10^{-23} of a single second. It sounds like the Marx brothers.

These are the tiny measures of time required in quantum mechanics and particle studies so let us not stop until we reach some final destination of extreme smallness of time and distance.

In pushing to the full hypothetical limit we finally arrive at the Planck Time Unit as the shortest measure of time bearing any meaning in physics, equal to 5.39 x 10^{-44} seconds. A Planck time is the duration a photon travelling at c in vacuum requires in covering a distance equal to a Planck length.

As for this minimum meaningful distance, the Planck length is defined from three fundamental constants, considerations we will examine more closely. Apart from flitting particles in great collider machines, these extreme micro units of time are invoked in the exegesis of the most initial phases of the Big Bang, describing its earliest periods and some incredibly compressed and brisk development. A lot happened in the first second. Indeed, we envision six whole epochs including the first emergence from the Planck epoch into the grand unification epoch between 1×10^{-43} and 1×10^{-36} s.

In the opposite direction, colossally large spans of time and distance are integral to the unfolding story of the universe, including that impressive conclusion that the cosmos is a reliable 13.7 Ga or nearly fourteen billion years old. There are several strong lines of reasoning supporting that the Big Bang was that far in the past. (See Weintraub's *How Old Is The Universe?*)

Prima facie, this is all the time there is but the origin and expansion of the universe is emphatically not that simple or linear. The sublime magnitudes of cosmological time are positively *not* just a passive background of what amount has elapsed since the birth of the universe. By "now" it certainly extends further than 13.7 billion light years in radius and as we have lately found, the rate of expansion has probably not been uniform ever since. It bears huge implications that the rate of expansion has actually increased.

Perhaps it is best to start with the original question, *viz.* why *is* there something rather than nothing?

Could everything have originated in the grandest quantum flux providing the biggest ever free lunch? What caused the Big Bang? What existed before? Are there veritable multiverses? Proponents of M theory would certainly have it that way and one key inference of String Theory includes the lengths and conditions in which particles could emerge.

Even the constraint of a single cosmos demands the visible components to be an estimated mere 4% of the total. Dark energy and dark matter are far more prevalent.

Major considerations of curved space-time, dynamic early inflation and the time spans required by signals to reach us strongly apply. In the oeuvre of the observable as opposed to the visible cosmos, there are signals that can never reach here due to the furthermost expansion. We will strain our brains appropriately in the next chapter.

It is time to get down to serious business and consider the birth and lat-er, some conceivable long-term outcomes of the universe. Let us attempt to begin before the beginning. The subject is engaging but it comes with an intellectual warning.

What was going on before the cosmos began, this mysterious time before time began? Like all successful epics where the actors have moved on, we must consider writing a prequel. In the theater of pre Big Bang cosmology, physics unavoidably meets philosophy and there is an apparent merger of astrophysics and metaphysics.

What guidance might our intellectual heritage provide here? Firstly, we consult one illustrious ancient Greek we know well. We then refer to the theological orthodoxy that long grounded Western thought and proceed to consult the great thinker of the European Enlightenment.

In order, we entertain Aristotle's view that the universe has always ex-isted, perfect and unchanging. Then in Semitic religion the eternal deity perpetrated a deliberate great act of creation and by Christian *theologos* God has always existed. In the 5th century AD Augustine suggested that time it-self came into being with creation and this deduction holds in the standard model for the Big Bang. Early Church fathers including Origen, Theophilus and Justin Martyr believed that matter was pre existent and it was Platonic influence to pose that God brought cosmos out of chaos.

For our spokesman of the Enlightenment, it was Kant who asked, if the universe had a beginning, why did it wait an infinite amount of time before it began? His antithesis was that if the universe had existed forever, why did it take an infinite amount of time to reach its present stage?

As we approach what is possibly the biggest question our enterprising minds can pose, we shall be highly selective. We shall give a nod to the mav-erick Peter Lynds and his physics paper addressing Kant's *Critique of Pure Rea-son* entitled *On a Finite Universe With no Beginning or End.*

Kant argued that the notions of a universe having a beginning and stretching back infinitely in time, versus it had a beginning at some finite point in the past, were equally self-contradictory. To Kant, this idea — that time could stretch back forever — was absurd. On the other hand, if the universe did have a beginning, what happened before to cause it? And what before that? These are indeed "timeless mysteries" that we may puzzle over forever; the answers are not yet within the grasp of science.

For us, the most fundamental question is this: What caused the Big Bang and where did the point source that became this unrivalled explosion of energy and matter come from? If that is not enough, there is the corollary of the requirement that everything observable and beyond derives from a tiny particle of infinite density and geometric curvature.

If the cosmos had a discrete beginning, then the hypothesis of an eternal universe must be false. What went on before that is like asking what lies south of the extreme south polar point? Like zero as denominator, it is not defined.

We must be content for now to hold that time and space derived from the Big Bang and there was no previous era. It was a day without a yesterday. Enter the paradox of infinite regression. This still boggles the mind of the most seasoned scientific theorist. Reason breaks down in the face of such a perfect paradox. It is a zero data conundrum and philosophical brick wall. A variation, equally unprovable by today's science, is the idea of a Big Crunch where we postulate an oscillating or pulsating universe in which the Big Time is cyclic and a series of collapses and reexpansions is the grand case. There are several ambitious books on the subject (Gasparini's *The Universe Before The Big Bang* is recommended, and *Cycles Of Time* by Penrose), yet I feel I learned little from them when it comes to the crunch.

Let us steer away from a philosophical approach for now. I most reluctantly consider that, push them as we do, there are limits to human understanding, including the concept of infinity and any comprehensible domain of the pre Big Bang.

Differently, $\sqrt{2}$ or an exact value for π are lengthy arithmetic projects. Seeking complete answers to such ostensibly simple questions apparently involve unending calculations. Their decimal representation neither end nor settle into repeating patterns. Infinite series are frequently abroad in mathematics whilst imaginary or i numbers like $\sqrt{-9} = 3i$ are even more fun.

We shall now engage the proven principles of special relativity in time dilation. It is an imposing truth that time flows at different rates under extreme circumstances such as relativistic speeds and gargantuan gravitational fields that locally distort spacetime. This occurs around neutron stars and black holes and we will shortly examine two more minor but significant cases in our cosmic back yard.

Time dilation strictly excludes astronauts orbiting at speeds that are very low in relation to c. The rate of motion of the International Space Station is a mere 0.00003% c. This being said, astronauts on orbital flights age 3 nanoseconds less for every second passing on the ground, accumulating to 3.8 seconds over a year.

Rest assured that our capabilities in spaceflight so far are ponderously slower than where time dilation's effects seriously kick in. Some very significant % of c is required for the slowing of time to noticeably ensue compared to the outside observer.

During everyday motions this does actually take place by entirely negligible degrees. As for when it does start to matter aboard our super duper starship, be advised that time dilation does not bear a linear relationship to increasing velocity. More importantly, in approaching relativistic speeds the increasing energy put into acceleration fosters effects other than merely speeding the vehicle up.

It promotes an increase of mass and decrease in volume in addition to the slowing of time on board compared to the rate passing at mission control. It only gets exponentially serious as we accelerate and add further 9's to the 99.99999 % c. The speed of light cannot be achieved for any particle with mass but there is no theoretical limit to the closeness to c that it may be accelerated to or its possession of kinetic energy. We will take a few thought experiments aboard starships in Chapter 3.

Until then, another brief foray into the story of physics is called for. Unlike sound, the speed of light proved too fast and hard to measure by simple terrestrial means. Galileo and others had made some bold but inadequate attempts, leading Descartes to conclude that it was simultaneous and infinite. As often with science the first real inkling was accidental. In 1676 at the Paris Observatory a speed for light was indicated to Rømer by the shifting timings of the moons of Jupiter as caused by the changing distances of Earth to Jupiter and the varying times of the arrival of the light.

As for little "c" for the speed of light Weber and Kohlrausch introduced the term from the Latin *celeritas* or truth in 1856 and Einstein later utilized it.

In the 1860s Maxwell used "V" for its velocity in his equations for classical electromagnetism and a refined quantitative value comes out.

So, having established that there was no questioning of time's absolute passage due to our natural unfamiliarity with c, super gravity or warped spacetime, this explains something more in general terms.

The reason why parts of quantum theory and relativity defy common sense is that they govern conditions wholly different to normal experience. We are not suitably prepared for it with the mindsets adopted in the very process of reasoning. We require uncommon sense.

We seem to be getting somewhere with space, time and energy. We could conclude that the universe in no more than the concordance of matter and energy with a defined relationship of $E = mc^2$, Einstein's greatest hit. Throw in the laws of thermodynamics, six vital constants and some inventive protein chains, simmer for 14 billion years and again.... voilà!

Having found it so integral, what then is matter? As usual, the Greek atomists Lucretius and Demosthenes are our first theorists, with *atomos* for "too small to be cut" describing the imagined basic particles of composition.

Such miniature building blocks of the material world have long been deduced to exist. However, they were supposed to be simple things of fundamental construction, which proves far from the case. The chemical definition of an element as a pure and individual single substance that cannot be broken down into anything simpler in combination still holds.

However, individual atoms bear elaborate structures and naturally occurring isotopes. One way by which an element is distinguished is by its atomic number, the number of protons in its nucleus. The Periodic Table is another masterpiece of the succinct. It predicted undiscovered members of the naturally occurring elements.

Following Rutherford's experiments first discovering an atomic nucleus in 1911, atoms soon revealed composition and organization both within themselves and discrete orbits of surrounding electrons.

Atomic substructure, energy levels and the strange, strange worlds of quantum theory have since been revealed, seemingly without end. Here derives the wisecrack that one year we should award the Nobel Prize to someone who has *not* discovered a new particle lately (Weinberg). We should add that in their early theoretical developments atomic structure and relativity as worldviews drew critics and active detractors.

After considerable debate, probabilities trumped certainties. Originally, Einstein was averse to anything offending strict determinism and had assumed a static non-expanding universe over fixed time as he explored the electrodynamics of moving bodies. Classical physics knew no other paths of enquiry.

Maxwell and Newton were among his historical mentors but their pictures on the wall of his office gave no cause to think otherwise. On the mathematical inequalities asserting a fundamental limit to the precision with which certain pairs of physical properties of a particle can be known simultaneously, Heisenberg was proved correct over Einstein with the Uncertainty Principle. Well, probably. In summary, determinism was being superseded by relativity and quantum theory.

Coming fully up to date, one important new issue concerns the Higgs boson particle, possibly with more than one kind in the echelons of extreme elementary particles. The Higgs field is the simplest of proposed mechanisms for the breaking of the electroweak symmetry and the means by which elementary particles acquire mass.

I think "God particle" is a ludicrous term, but we need a catchy name to make attractive headlines. "Smallest Thing Found" definitely lacks appeal to intellectual tourists looking for a good read in a new magazine article.

New developments in dark energy, M theory, particle physics and SETI are subjects worth revisiting once in a while to see what has been decided of late. A breakthrough with the latter is long overdue and, like walking on Mars, I wonder if any of us will live to see it.

As noted, results from the Planck satellite mission now suggests a little lumpier initial universe of slightly greater age with a little less vacuum energy than previously supposed. We emphasize the suggestion that dark energy somehow took over and refurbished the expansion of the universe about 5 Ga ago. This is a real possibility. Separately, The Laser Interferometer Gravitational-Wave Observatory's recent detection of gravitational waves from a double black hole merger was long awaited.

Neither receding galaxies nor spinning electrons are inherently obvious, despite the best-applied efforts of our hard thinking Hellenic forbears or the celebrated proponents of the Age of Reason. However enlightened the minds applied to the visualization of the great and small, things as tiny as subatomic particles and as big as galaxies are not on direct show.

What of the inner designs and configurations of atoms such as neutrons, protons and electron shells? Evaluated in conjunction to the expanding universe it seems an invitation to speculate on dual infinities spreading both inward and outward. This challenges all logic but appears the case to all our applied scrutiny so far. There is no end in sight in either direction, providing unlimited future investigations in diametrically opposed scales. I guess we do need those special units.

Time and distance is a two-way colossus, ranging as big and small as you like. It is unwise to invoke limits just as the resolution of one mystery opens the doors to greater puzzles. There are fascinating interfaces of particle physics with early Big Bang cosmology and now of course, String Theory and the speculations of entire other universes.

Why limit to a mere four dimensions? Greene's *Fabric Of the Cosmos* is one of the best introductory studies here. Dare to swim in quark-infested waters and speculate that far. In science we have a distaste for the idea that there are things unknowable on principle. However, both the Uncertainty Principle hijacks strict determinism and separately, there are limits to an *observable* universe.

A word of warning is due. We should not pass over the profound problem that the physics of the very large and very small are incompatible in our understanding, placing a serious conundrum on any fusion of quantum physics and general relativity. Here is the prime obstruction to constructing any Grand Unified Theory, Theory Of Everything or the Unified Field Theory that Einstein pursued in his later work. It remains unfulfilled; indeed such a thing remains defiantly elusive.

The very meaning of "now" is subject to interpretation because even light takes significant time to cross the distances between stars and galaxies. I live in awe that such speeds are so very common and that even the ultimate velocity is demonstrably finite. As a feature of the propagation of the electromagnetic spectrum we reiterate that c is built into the fabric of spacetime — however the latter is stretched, compressed or sped across.

All received signals are from some depth of the past, commencing with the one and a third seconds to arrive here from the moon and just over eight minutes from the sun. There was an apparent hesitancy in the voice exchanges from astronauts at the moon due to the half million miles there and back. Jupiter is seen a varying forty minutes in the past and placing the boundary of the solar system as the orbits of Pluto and inner Kuiper belt it is roughly five light hours in diameter.

Bradley had attempted to measure the parallax of stars in 1739 obtaining worthwhile consolation prizes with the discoveries of the aberration of light and the nutation of the earth's axis. Within a century the authentic plumbing of an interstellar distance became something of a scientific race and Bessel is credited with winning it in 1848 (see Hirshfeld's *Parallax.*)

For the star 61 Cygni whose considerable proper motion suggested relative proximity Bessell successfully measured an annual parallax of 0.314 arcsecond and published the results. Given the diameter of the earth's orbit this indicated 10.3 light years, actually a 9.8 % underestimate, all told. For all previous ages the distance to the stars had been an imponderable and semi-mystic question.

Who can comprehend 186,282 miles a second, 670,616,629 miles an hour or 5,865,696,000,000 miles covered over a year by a ray of light? This is almost 10^{12} or 6 million million miles, invoking the "trillion." No single individual invented the term "light year" and even the least distant stars require such custom built large units. I always observe 61 Cygni on a clear summer's night.

A parsec is the distance at which one astronomical unit subtends an angle of one second of arc, equal to 3.26 light years in applied astrometry, which is concerned with the motions, positions and magnitudes of stars. On principle every star reveals some parallax as the earth orbits the sun, a fact sadly hidden from Tycho Brahe's reasoning. The Tychonic system where the planets orbit the sun and the whole assembly goes round the earth was a forced combination of geocentrism and Copernican heliocentrism and drew little support.

Herbert Hall Turner suggested 'parsec" in 1913 as a portmanteau of "parallax of one arc second." As for the nearest star after the sun, Alpha Centauri lies 4.2 light years away. Its component white dwarf Proxima is currently the closest member of the system with an extensive orbit around the parent A and B stars.

How far have we reliably seen by now? One is often asked this question.

Images from the Hubble Ultra Deep Field as long photographic exposures have broken the distance record several times. This often hits the news. With red shifts equal to 8 this indicates a mere 600–900 Ma *after* the Big Bang which is a powerful advance on the more routine "ago." At an even greater redshift the galaxy GN-z11 may be observed from a mere 400 Ma post Big Bang.

These faintest and reddest objects are the oldest galaxies yet identified, lying at over 13 thousand million light years and significantly, only that few hundred million years *following* the Cosmic Inception. These are signals currently received from a time so remote that the universe was fundamentally younger and that things have significantly moved on in the interim. "Now" is not fixed at all.

A discrete and luminous event like a star exploding into a nova has happened, is happening or is yet to happen as a simple function of progressively greater distance from the source of the action. The nova was seen in the past, is currently on show or will eventually blaze into view to further flung star systems across a galaxy and beyond. The expanding spheres of radiation naturally proceeds at c, but like all waves or information conveyed, light takes significant time to cross the gulfs. This still applies at the fastest speed possible and time *not* passing by its own frame of reference. In crossing the solar system and reaching the first neighboring star, contrast five hours duration to just over four years light transit time.

There is no news, only history lessons of a greater or lesser extent. The stars you routinely see are a few tens, hundreds and in some instances a few thousand light years away. At 2 million light years the Andromeda galaxy is the furthest naked eye sight as shown by its inclusion in Al-Sufi's, *Book Of Fixed Stars* of 964 AD. Very good personal vision can discern the Triangulum spiral galaxy M 33 which is a bit further. Most impressively, Bode's galaxy M81 lies 12 million light years away and at visual magnitude 6.9 might be available to those with exceptional eyesight.

With deep sky observing, seeing an extra galactic object means telescopically capturing a much-travelled sample of its photons from millions of year ago and there are no exceptions to this profundity. It is the colossus of distance. Events are not seen to occur at the same time in any absolute sense but from different frames of reference. There is no absolute time. Indeed, there are no absolutes other than the velocity of light. Let us now look to the earliest proofs of relativity.

Einstein's general theory of relativity was first published in 1916. At the time it had no empirical basis but according to its author, the situation was nothing insurmountable. He saw three possibilities.

Firstly, Einstein held his theory capable of expounding a previous problem in celestial mechanics and planetary astronomy. An adjoining paper successfully had earlier tackled the long-standing issue concerning the secular advance of the perihelion position of Mercury's orbit, an anomaly in celestial mechanics, which had remained inexplicable since the work of Le Verrier in 1859.

In summary, gravitation is measurably mediated by the curvature of spacetime and the pull of the sun has a measurable influence on the planet's regular closest point, specifically 45 arc seconds a century. This relativistic explanation for the anomaly of motion for the planet Mercury superseded a tenet of strict Newtonian determinism. For the first time the latter was revealed as insufficient, forming a significant point of departure from classical to relativistic physics.

Secondly, there was the proposed bending of starlight surrounding a totally eclipsed sun. By its mass, the sun's gravity should permanently cause a minor distortion in the spacetime of its environs that is indirectly observable under such special conditions. Gravity can bend light and the sun is clearly the largest gravitational filed in our vicinity. According to theory its attraction should measurably affect the paths of sunlight around its rim and alter the observed position of surrounding stars compared to their precisely charted positions.

The third hypothetical test of relativity involved speculations on gravitational redshift, something he felt that the instrumentation of the era was not up to. Attempting this by the spectrum of the white dwarf star Sirius B was inconclusive in 1925 but we have since got to better grips with this. Commencing with Popper's measurement of a 21 km/sec gravitational redshift for 40 Eridani B in 1954.

For a proposed demonstration Einstein's confidence ran high for when a total eclipse of the sun next presented. An aborted attempt had been made a few years earlier when his overly obliging assistant Freundlich mounted an eclipse expedition to the Crimea in August 1914. Unfortunately, they ran into serious earthbound difficulties. With the outbreak of war with Russia, the German nationals could not have found a worse time or place to attempt the observation, not that there is ever any choice for where or when the track of totality touches the earth's surface. The group was interned as spies and their equipment seized.

In November 1919, the Englishman Eddington and collaborators were to have much better luck with this repeatable experiment and falsifiable conclusion — as science purists demand. They attempted to photograph the effects measurable about a totally eclipsed sun when specific stars immediately around its rim become visible for a few precious minutes.

On the island of Principe off the west coast of Africa and elsewhere in Brazil his teams attained such photographic plates. Back in London with them the predicted effects were there and the results launched Einstein into international celebrity. It was a happy event for post war Europe and played out on the world's stage of publicity. Later, Eddington further confirmed the theory's success in fully explaining the perihelion shift of Mercury, something Freundlich had earlier asserted in a more academic sortie. Its precise amount predicted by relativity indeed corrected the small quantity unexplained by Newtonian theory.

A journalist asked Einstein what he would do if the gravitational effect on starlight had failed to match his theory. He was confident, asserting that it would be a practical failure of observational technique if they were not apparent. He replied: "Then I would feel sorry for the good Lord. The theory is correct."

A Page In Earth History

Let us hold the winged horses of frontier science by the reins of humanism a while longer. Soon we shall ride them into the far reaches of spacetime and ruminate on the cosmic evolution that leads to us.

We saw that the veneer of recorded history is based on surviving texts and archaeology. It is the first step in rationalizing levels of time on our own terms. We saw that positing some 2 Ma of age for mankind places us as a recent occurrence in the fullness of Earth history's 4.6 Ga. It provides an important perspective when we compare ourselves to the age of our birthplace. By any comparison we are new kids on the block.

What facts are unequivocally established by fossil and geological evidence? Firstly, strange looking fossils can be relatively thinly laid or even as outcrops in plain view. It is easy to see such remnants of creatures that no longer exist and earlier versions of some that do. Deductively, the kinds of animals we see around us have not been around for long as species.

Man was certainly not coeval with the dinosaurs, the most famous of extinct creatures whose fossilized remains have that definite point of cutoff in the geological record. We shall look at their passing as a tool to better understand our time and place in context. As the rock stars of extinction, what caused their demise?

It was the most recent single event of mass extinction, if not the largest of a series that have occurred perhaps five times over known Earth history. A worse one occurred 250 Ma ago with the Permian-Triassic event that destroyed perhaps 96% of marine and 70% of all land life. The K-Pg event was merely the most recent.

Welcome to the planet of the dinosaurs. Once upon a time, some seriously big lizards lived on a small rocky planet going round a regular star. They were lots of them of different shapes and sizes and some were truly imposing; definitely the biggest creatures that world ever gave rise to.

Then in one catastrophic moment an asteroid struck the planet, killing off a whole lot of life. It destructively altered the biosphere and severely damaged the environment by direct and indirect means. It was bad news in the short term but things recovered remarkably well over the time that ensued. Despite a massive setback, new directions and evolutionary paths were slowly adopted by the surviving life forms — though creatures were never to be generally that big again. Eventually, conditions proved much to the advantage of some particularly smart monkeys, later giving rise to a veritable planet of the apes. Whilst this might sound like a corny science fiction plot, this definitively occurred here 65 Ma ago. Let us paint our best picture of the events.

On the immediate impact of the intruding cosmic body the atmospheric shockwaves no doubt encompassed the world several times. The shock must have reverberated through land, sea and air on a global scale. The energy released probably precipitated heat and fires lasting months on end to consume whole sections of forests and vegetation. Possibly the conflagrations spread about whole continents, combining several forces of destruction.

Then there were drastic falls in worldwide temperatures, caused by the vast quantities of superheated rock and biomass flung into the atmosphere abruptly blocking the heat of the sun. Everywhere it became suddenly colder and in a dispassionate assessment, most parts of the food chain collapsed overnight, freezing and starving much life that had survived. (The food chain is defined as the hierarchical series of organisms each dependent on the next as a source of food. It commences with producer organisms and weaves its carnivorous or herbivorous great chain up to apex predators.)

From the evidence of fossil pollen and leaves, photosynthesis in plants and plankton was abruptly curtailed as the very atmosphere became impervious and possibly poisonous. It is estimated that 57% of plant species in North America were wiped out by this means. We think that the dust did not last long but sulfate aerosols made the longer term cooling effect because they last longer in the atmosphere. Following the impact gases bearing sulfur evaporated to form them causing the average global temperature to suddenly drop probably below freezing for perhaps three years.

The recovery period may have been about three decades as photosynthesis was able to resume, curtailing a temporary proliferation of fungi that do not employ the process. The impact winter was devastating. In the big picture it could be asked how close did things come to wiping out all life but studies suggest no, not that close at all. It is also remarkable how bio diversification got previously carried away to make animals so very large. For example, how much of their waking time did a big dinosaur spend foraging and hunting for food?

For diversity and individual sizes, the dinosaurs existed in profusion on every continent including a limited presence in Antarctica. No tetrapods

weighing more than 55 lbs survived, although on evidence, amphibious creatures did better in survival rates.

The demise of the dinosaur and so much else is fairly well understood
by now. Indeed it is a triumph of scientific investigation to have unraveled
its causes, the evidence being conclusive for the exact time and place. The
Alverez hypothesis was first put forward in the 1980s.

The blow from space was struck on what is now the Yucatan peninsula
in present day Mexico. Here lies the large and mostly submerged Chicxulub
crater, the scar left from its original descent. Today the remnant crater is
some 110 miles in diameter and 12 miles in depth, slimmed down and eroded
by subsequent geophysical forces. The impacting body was relatively small,
perhaps some ten miles in girth and depending on speed, density and other
factors it packed the energy of thousands of nuclear bombs.

At the time a telltale global quantity of iridium was laid down to form a
specific geological boundary of anomalously high concentrations in a worldwide thin layer of clay. It measures about one hundred times the normal
traces of iridium, which is comparatively rare in the earth's crust and was
therefore of extraterrestrial import in one go. The accumulated evidence of
shocked quartz, gravity anomalies and surroundings tektites further indicate where the impact occurred. Possibly, the impacting body could have
been cometary.

As for these huge and bizarre creatures the largest sauropod dinosaur
measured 120 feet from nose to tail with an estimated body weight of 100
tons. A single vertebra of the recently discovered Argentinosaurus was 4 feet
in girth. Could they have grown more diverse, populous and still larger as
individuals, or had they reached some operational limit for an environment
to sustain them?

I think the latter is the more likely scenario because the resources are
finite but what might their unaffected future have held? What would their
descendants be like by now? We deduce they lacked the dexterity, communication and brainpower to have developed structured social organizations
or the application of technology.

The asteroid or comet could have missed and the terrestrial career of life
would have been radically different under conditions allowing the great lizards to endure. Not only would the rise of large mammals including man have

been precluded, we could assume that *Tyrannosaurus rex* and friends would be alive and well to this day. Their sheer weight of numbers suggests this and T rex had probably the most powerful bite in biological history.

Instead, sudden doom from the skies ended their whole era. It was the most critical turning point in the geologically recent past and the evidence is widespread.

There were debatably one thousand dinosaur genera, again subject to serious uncertainties of classification. They are generally known by their genus rather than species name, the two main types being the lizard hipped "saurischia" tending to carnivorous feeding patterns and the birdlike hipped "ornithischia" which were mostly herbivorous. The giant sea reptiles may have given birth to live young rather than lay eggs and we surmise that the survival rate of eggs was naturally low despite burial before hatching.

How devoted they were as parents cannot be said, for example, did they build nests? Today we have fossilized egg survivors but embryos are very rare so we are still a long way from Jurassic Park or any bioengineering scheme of genetic reconstruction. If we ever gained that sort of capability there are far more important projects to address.

The chronologically earliest fossil evidence of them, laid down some 225 Ma ago attests to a dominion lasting some 160 Ma. Many uncertainties still exist including their hunting, social and mating patterns, also the color of their skins and the vocalizations they shrieked into the primeval night. The food chain must have been monstrous in action and we only assume it had struck a sustainable balance.

It is believed that some kinds of birds from the theropod group are their living descendants and decisively not the larger living lizards. Crocodiles belong to the same archosaur group as dinosaurs but are not directly linked. Non-avian dinosaurs were exterminated along with so much else, making many birds their actual heirs. Note they as proposed to be of saurischian ancestry.

In evolutionary theory the process of adaptive radiation allowed divergence into new forms and species to occupy the ruined ecological niches. This seems a rather bland act of description but there was certainly a major diversification among mammals in the new Paleogene geological period.

This most recent mass extinction occurred at the newly revised Cretaceous-Paleogene or K-Pg boundary. It was certainly severe and It was formerly known as the Cretaceous-Tertiary or K-T boundary. The layer now termed K-Pg is like a geochemical signature forming one specific calibration in the geological record.

We deduce this primarily from the iridium layer, which is unequivocally of extraterrestrial import. This new term K-Pg boundary requires explanation. A geological time chart might assist. Firstly, the "K" for Cretaceous comes from the original Latin *creta* as derived from *kreide*, the German word for chalk.

"Tertiary" has now been denigrated as a formal term for both a unit of time or rock. The impact ended the Cretaceous period and commenced the Paleogene period leading into the Neogene period that officially replaces the Tertiary period in the terminology. The asteroid strike is confidently dated at 65.5 +/- 0.3 Ma ago and fully seven epochs in our reckoning before *Homo sapiens* made any incremental appearance.

Here the student gains that exciting sense of temporal proportion not exclusively reserved for the astronomical sciences. Earth history and geophysics partly serves just as well, incorporating the latest one third of full cosmic history.

Equally, two thirds of the time elapsed since the Big Bang had passed before the sun formed and these facts alone provide a vital perspective. We can confidently estimate when the sun, the planets, life and man all first arose although much of the processes are still unplumbed.

A whole new epoch, period and era was kicked off that 65 Ma past. Such was the significance for the transition, singularly due to that specific bolt from the blue. In a single moment one inert and primeval rock left wandering from the early solar system had shattered 3.8 Ga of biological progress.

Restless Earth

We shall now consider a little of the earth's immediate face and active past. The shift and positioning of continents and oceans amidst varying climates, transient ice ages and several mass extinctions are revealed as we attempt to dissect the great timespans. Allowing 4.6 Ga in total age, the most recent passage of 65 Ma covers but 0.014% of complete Earth history.

We estimate that five major and numerous minor glaciations occurred in the last 2.5 Ga, possibly from different causes. The most severe episode may have produced the so-called Snowball Earth lasting from 850 to 630 Ma ago when ice reached the equator. Things tend to get a little hazy in approaching the remoter past of Earth history including the comparatively early emergence of life. On evidence, this was a little less than 4 Ga ago and certainly in the Precambrian era. In proportion, the first appearance of living cells was less than a half billion years after the formation of the earth; a highly significant find when viewed in proportion.

Much is directly on show in the strata of rocks as the repositories of fossils, driving paleontology as the study of past life. Unlike some branches of research, the full natural laboratory of the uppermost physical world is directly available to hands-on probing and demands more modest budgets than space missions.

The Latin *fossus* means "having been dug up," with the arbitrary minimum limit of 10,000 years of age defining fossilization. Yet it is a literal mere scratching of the surface, our knowledge of the innermost makeup of the planet being limited. Ideas about the earth bearing a structure of outermost crust phasing into an underlying mantle leading down to liquid outer and solid inner cores are based on seismology and modeling, not direct sampling.

Fossils themselves were unexplained curiosities for many years, being common to the point of occasionally obvious. Long before elephants appeared they were the (fragmented) brontosaurus in the room. The ancient Chinese regarded dinosaur fossils as dragons from the past, which is figuratively true. They had uncovered many a fossilized skeleton in their mining projects.

Certain types of fossils are associated with specific rock strata and their relative age by graduated depth in the stratigraphic column became apparent to the more methodical investigators of the 19th century. Modern means of radiometric dating allow reliable absolute ages to be determined. The decay rates of natural carbon isotopes can be assessed for former organic material. It allows us to approximate within limits how long something once living has been dead. Related techniques of dating may be applied by naturally decaying isotopes to assess surrounding the rock layers of fossils.

Aristotle correctly associated fossils with life from the past and the Persian scholar Avicenna (*c.* AD 980–1037) successfully mused on the "mineralizing and petrifying virtues which arises." (He is also quoted as preferring a short, broad life to a long but narrow one. This much he was granted.)

Albert of Saxony expanded a similar view of fossils in the 14th century. Leonardo Da Vinci concurred with Aristotle, posing questions concerning the vestiges of strange marine specimens found in places where there are no longer bodies of water. This too came to challenge the established beliefs of Noah's worldwide deluge and we encounter a certain pattern of frontier science rubbing up against religious orthodoxy.

Fossils range from biochemical traces and microscopic single cells to large dinosaur bones and whole petrified trees weighing tons. Soft tissue remnants are comprehensibly rare in generally not being preserved. The chemical processes generally operate on the parts that were partly mineralized during life, like the bones and teeth of vertebrates and exoskeletons of invertebrates. However, eggs, feces and external marks like footprints of great reptiles have been occasionally preserved, a set of the latter is attributed to *T. rex.* taking a stroll.

The fossils identified (with a few embarrassing goofs) then painstakingly reassembled as they were in life reflect the cold glory of vanished creatures of strange forms and nightmarish sizes. When my son was small I promised him the sight of "real dinosaurs" in London's Natural History Museum and he was not disappointed. A childlike wonder is a good thing in natural science and the balance of fascination with mature skepticism is even better. Call to mind the sights of whole herds and flocks of dinosaurs traversing the plains, seas and skies of the Mesozoic era. Compared to current times, Earth was an alien world to itself.

Those marine fossils found up present day mountains were strong hints that the locations of sea and land change over time. There are plenty of marine dinosaurs in what became the American Midwest. Sedimentary strata bearing fossils can also be raised to greater physical heights by mountain building.

Certainly, surface conditions on Earth have shifted enormously time and again. This has involved changes to the positions of whole seas and landmasses and over lengthier Earth history, the chemical make-up of the atmosphere. Bacteria and plants set the stage for life on land.

Our world continues to change. The Earth is not an inert object but a series of dynamic, interrelated and ongoing processes, more active than any other world in the solar system with the qualified exception of Jupiter's moon Io. (The important lesson here is there are heat sources other than the sun. The depths of the solar system are not necessarily mean abject cold.

We are ensconced on what are actually only limited regions of climatically comfortable land, surrounded by a life-supporting and protective atmospheric envelope. In ascending mountains we can directly climb above levels of breathable air and the comfortable temperatures of the lower troposphere where we dwell. A few miles cannot be considered far, even to Flintstone man. Just consider such modest distances upward in direction, where the air soon becomes tenuous to breathe and conditions dangerously cold. We seldom reflect how precariously we exist on the thin surface of an active world that has undergone massive changes over time. This serves a dire warning about the conditions we consider so fixed and secure.

What exactly is taking place that can be discerned by simple physical geography? Take a look at a simple map of the world.

Primarily, continental drift and plate tectonics continue to sculpt the face of the world and ocean floors. These are still active as earthquakes, tsunamis and volcanic eruptions graphically illustrate when something builds up and suddenly gives way. Some parts of the continental divides are evident on the surface and globally, there are at least two hundred seismic occurrences going on at any moment. The positions of the continents and seas are permanently on the move with no final destination.

More transient but equally menacing to us are the meteorological fluxes of the atmosphere causing hurricanes, deluges and other violent weather. (For clarity, meteorology is the state of the lower atmosphere and climate the average of weather conditions over time.)

Whilst these are nothing new, it is now wedded to greater extremes of temperature by overall global warming and the impending rises of sea levels. Polar ice is melting. I will not waste ink that this is other than an urgent matter.

When a seismic flutter affects the uppermost crust to cause an earth-quake or the sea floor moves to propagate tsunamis, they inevitably appear big to the point of apocalyptic to us. Solid land moving like waves of water is a strange sight and a most eerie sound. Crashing buildings and giant waves are terrifying. It is impossible to view such shakeups as minor when there are massive losses of human life. These inconveniences are something we have to live with and occasionally die by.

With all respect to those killed and the families bereaved, neither Hurricane Katrina nor the Japanese tsunamis were anything significant in Earth history compared to the K– Pg extinction event. As geophysical processes the seismic shocks and tidal waves of recent years must be seen in proportion. They are mild compared to what could take place and we permanently worry about the "big one" *i.e.,* a major earthquake regarded as inevitable on America's west coast.

The recent disasters proclaim the vulnerability of man, his existence contingent on physical conditions on which he holds limited understanding and zero mastery. Our best abilities preclude any actual control and only too swiftly conditions can turn deadly. They translate into major tragedies due to the fragility of our condition.

But at least our knowledge of the actions of the global tectonic plates and seismological surveys of the world's hot and mobile regions give better explanations than angry gods out to get us. Seismic disasters have afflicted mankind for so long that they entered his mythology yet the state of the lithosphere on the move was not understood until the 1950s. (*Lithos* is Greek for rocky and *spharia* for sphere.)

A cursory glance at the present face of the globe resembles a giant jigsaw with more sea than land and an independent frozen continent at the far south. The ratios are about 70% of water to 30% land with immense subterranean water present beneath the "solid" land. Here is something positive from these statistics. There is a greater volume of living space in the seas than the land by a factor of three hundred.

There is enormous room for cities and settlements under the oceans should we ever explore such ambitious vistas of metropolitan engineering. Construction of such abodes on the most stable continental shelves would be safer, in strict avoidance of the active mid ocean ridges or where the crust is thinnest.

Perhaps a fully sane and technologically accomplished future society could address these possibilities if we shy away from big colonies on the moon or Mars. Why not develop both? We are already facing a crisis of over population and something has to be done if we are to survive another century.

Tectonism is the set of processes going on within the large-scale structure of the earth's crust — from the Latin *tectonicus* for building. The tectonic plates are driven by the released energies of radiometric decay setting up currents in the mantle and deeper, there is possibly a contribution from trapped primordial heat. These geophysical actions are unique to our world in both global scale and full continuance into the present cosmogonic era. Here is a mystery in itself within comparative planetology but it is encouraging to know enough of the structure of neighbor worlds to allow comparisons. For so long the planets were merely telescopic images.

Why is tectonism still prevalent to Earth alone? Why do the crust and upper mantle remain mobile after so long? Here are two descriptions.

Firstly, the entire hard and rigid upper layer comprised of worldwide tectonic plates is a uniquely ongoing big feature here. Secondly, we are probably the sole world retaining an active and globally sized asthenosphere.

This latter lies 50–120 miles below the surface between the lower crust and the upper mantle and forms a hotter, weaker and lower section. Its lower boundaries are not well defined and at the mid ocean ridges the transition lithosphere-asthenosphere is only a few miles beneath the sea floor. The lithosphere remains rigid and tends to deform in brittle fashion and the asthenosphere (Greek *asthenes* for weak) accommodates strain by more plastic deformation.

Sea floor spreading, plates actively colliding or sliding along subduction zones can give visible jolts and emanate waves as dramatic earthquakes. The Indian subcontinent continues to move a little more up into Asia but recently delivered a significant jolt, felt on Mt Everest itself. This explains why the Himalayas are the world's tallest mountain range but regrettably the events of 2015 wrought havoc and suffering across Nepal.

Somehow, Earth's processes remain strongly more active than on any of the inner planets or myriad moons of the outer planets that have rocky outer compositions and differentiated full makeups. In principle, many are similar in terms of crusts, mantles and cores and we now find that the larger asteroids are differentiated in these ways. Here, the single supercontinent Pangaea probably emerged in the late Paleozoic and early Mesozoic eras about 300 Ma ago and only began to drift towards the present configuration a mere 200 Ma past. This is emphatically recent and of course, ongoing. (Greek *pan* means entire and *Gaia* signifies Earth.)

Widening our scope, Venus might possibly have a few active volcanoes on its hot surface beneath its dense atmosphere. Mercury and Mars are seismically quiet as far as we know with negligible and thin atmospheres respectively and the internal structures of the gas and ice giant worlds of the outer solar system remain serious known unknowns. Perhaps the great atmospheres of Jupiter and the others transition downward from gaseous to solid conditions through warmer liquid states rather than the abrupt atmosphere-to-surface or ocean situations pertaining to Venus, Earth and Mars. There is an anomalous amount of water on Earth that in origin cannot be fully explained. Temperatures allowing H_2O to generally exist as liquid and endure does not explain how so much got here in the first place.

Casting about the other planets, some lithospheric movements probably occurred on the Jupiter moons Ganymede and Europa. Their actions are apparent on the surfaces. Volcanic and tectonic processes are also evident on the surface of Saturn's Enceladus but they too are from the past having long reached slow or stationary. Titan, also in orbit about the ringed planet retains an appreciable atmosphere and possibly the results of some lithospheric movements.

Neptune's moon Triton displays geysers of frozen nitrogen visible as plumes but these may be atmospheric effects In this case we meet a dust devil hypothesis rather than cryovolcanos of water, methane or nitrogen but there is certainly some contemporary activity of limited scope. We have mentioned the possible subterranean ocean of Europa and the speculation of organic presence there.

Then there is the remarkable case of Io, slightly larger than our moon and one of the four principal Galilean satellites of Jupiter. It is the most active world of all with gravitational interaction acting as the primary energy source for the moon's volcanic action and continual hot resurfacing. Jupiter gravitationally kneads the pizza planet and serves it fresh like no other hot spot in town.

Its volcanism was one of the breakthrough discoveries during the Voyager encounters with Jupiter and the previously known thin sulfur ring along its orbit might have dropped a hint of what was going on. It is the densest body in the solar system, followed by our own slightly smaller moon. Its full seismic activity is a remarkable exception and its causes are entirely different cause and none of them are an analog of the tectonism still working on Earth.

Deeper down in the earth's lower mantle, the liquid outer core combined with the Coriolis Effect gives rise to the magnetic field. Its existence promotes a protection from dangerous solar and cosmic rays. This remains one of the great factors operating unseen although aurorae give a lovely display of dancing celestial curtains of light near the earth's magnetic poles. The sight of the solar wind interacting with the ionosphere is a wonderful sight and even better from orbit via the ISS web cam.

The magnetosphere is silently protective to cells, genes and organisms as the outermost and unseen component of the biosphere. The moon maintains our axial alignment, further contributing to stable conditions and the model for the solid inner core of the earth includes a diameter of some 760 miles and a temperature of 5,700 K, comparable to the surface of the sun. We also posit that it still spins faster than the more outer regions.

The general rotation of the earth continues to gradually slow, increasing the length of the day. The figures of a retardation of 1.7 milliseconds per century over the last 2,700 years suggest that the primordial Earth may have span in as little as six hours. Tidal drag also moves the moon further away, expanding the size of the lunar orbit by 3.8 centimeters a year.

In summary, everything here is the latest and transient outcome of prodigious geophysical and biological actions including the gaseous cycles accompanying the actions of plants and creatures. Antarctica was always separate as a continent whilst the world once bore a single great ocean we name Panthalassa. Pangaea may have been preceded by earlier supercontinents.

Concerning fossils we must revisit a serious proviso to their study. We are merely aware of a limited range of specimens leaving remains that endured to be found and incorporated into our researches. These factors pres-

ent major constraints on the full story because it cannot be a fully informative representation by any interpolation of the available evidence. There are
many lapses of continuity in the record.

Which creatures progressed or survived to living genera and which
ones hit some evolutionary dead end to vanish forever (and above all *why*
this selectively took place) is often puzzling. Competition and adaption to
changing environments clearly play leading rôles in some huge survival of
the fittest by natural selection. Darwin was basically correct in these areas
and in the battle for survival has always been ferocious. Many species have
surrendered to extinction.

It is clear that great changes in biological form have taken place and
no doubt continue to do so. Meanwhile, the astonishing diversity of life
is demanding to fully classify, yet alone convincingly expound. There are
suggestions of microevolution being seen in action but the studies are not
conclusive.

The pace is certainly slower than anything man has been around long
enough to observe although such things of recent extinction as sabre
toothed tigers and mammoths might qualify that observation. Their demise
might be easy to explain. We may have been directly responsible for their
disappearance.

Evolution is any change across successive generations in the heritable
characteristics of biological populations. This is a good enough description
but it remains an elastic and broad term. The biodiversity we see today defies tidy explanation and remains inexplicable however much time it had to
unfurl.

It also means there are huge gaps in any prospective understanding of
paleontology. With more questions than answers, the wonders beneath our
feet could keep us humble enough before we look into space.

Who knows what kinds of fossils lie at deeper levels than we have visited
in detail? Resting far lower in the earth's crust or at greater depths beneath
the ocean floors than probed so far? There are certainly unknown living species in the oceans, occasionally glimpsed and starring in many a sailor's tale.
Not only did life begin in the sea, we are still 70% water with our bodies
acting as biological spacesuits.

The lesson is this: there are far more extinct than extant species and some departed leaving no hints of ever existing. The great Chain of Existence has many discontinuities and the earth's earliest stages of development are closed to direct enquiry. Mostly, continuing geophysical processes have consumed the evidence, destroying or rather reprocessing the uppermost material.

Here we meet a special known unknown before venturing into space to our closest neighbor and assuredly the first step to the stars. Circumstances allow a vastly better reach back in time with our only natural satellite and in order to gauge cosmic and geological time in comprehensible ways there is much to learn from the moon. We shall carefully sketch its motions, structure and history.

The Museum Next Door

Everyone knows the sight and loveliness of the moon. She wears a thousand different faces and poetic eulogies but were you aware that most of what we see is old even by the standards of our planetary system or that the lunar poles are probably the very coldest places in it?

The results of its turbulent history are clear with the simplest glance and quite unlike the earth, the major processes shaping its appearance have long since come to a close.

Let us commence with what is directly on show, starting with those flatter bluish zones covering a third of its face. In his telescopic forays Galileo suggested they were actual seas, calling them mare (pronounced *maray*, plural *maria*) but note they are entirely dry. Chemically, the moon might also be the driest place of all. From those first probes venturing out there it soon transpired that the lunar globe has distinct hemispheres in gross appearance and that the maria lie almost entirely on the side facing Earth. Maria cover about 16% of the entire surface.

Since their formation there has been only the endless arrival of cosmic dust, solar wind particles plus small and occasional larger meteors. Whilst not remaining entirely as formed, the moon ceased significant internal action since the eruption of the mare basalts 3.16–4.2 Ga ago, tending to the lower value. The trace hydrogen atmosphere is a good vacuum, rendering it entirely exposed to space.

Saturation cratering, plains and highland areas are clearly seen. It bears a most rugged appearance with the curve of the terminator showing the body to be spheroid and not disc-like in shape. Galileo applied crater from the Greek for "vessel." One can consult a moon map or better still, gaze at its surface to approximate some logical sequence of events.

Deductively, the craters appearing fresher and more prominent were made in the less distant past because they have not been dimmed by the space weathering of eons. Equally, the maria bear less quantities of them having not been in existence for so long. The more energetic collisions left central hillocks within the larger craters where the surface rebounded on being struck. Smaller craters tend to lack them or the graduated ramparts of walls or rock slumping from major impacts.

With higher telescopic magnifications, endless quantities of craters can be seen, including chains of them by presumed common origins. The observer sees ghost craters ruined by further poundings, rilles of collapsed lava tubes and long extinct lava flows making up the magnificent desolation. On the long running debates whether they are impact or volcanic the former won the day.

We see an unchanged scoreboard of multiple impacts and massive outflows by long dead volcanism. There are impressive vistas in lunar astronomy e.g. the terminator throwing features into relief as craters fill or empty with sunlight and mountaintops catch the first rays of the new rising sun. Such differences of relief are actually due to the sun's changing altitude over the moonscape and may be discerned over mere hours of close observation.

The moon does indeed rotate on its axis over the month. In doing so in practically the same time as it revolves about the earth it permanently keeps the same face turned earthward. It is often not appreciated that the other side receives equal amounts of sunlight and darkness over the month making "dark side" another serious misnomer. Please read "farside" in distinction to "nearside."

Such captured rotation exists with many other planetary satellites as a feature of long-term gravitational association. The New Horizons mission revealed Pluto and its moon Charon to be fully tidally locked with both bodies permanently keeping the same faces to each other. Their axes of spin are fully matched to a revolution period of 6½ days and Charon's diameter of 750 miles is fully ½ the size of Pluto. Compare this to the moon being just over ¼ of the diameter of the earth.

Uniquely for the Pluto–Charon system, the barycenter lies outside their bodies. The only other case of this in the solar system is the center of gravity for the sun and Jupiter, which lies outside the sun. Our satellite remains the largest in comparison to it primary among the classical planets and specifically the fifth greatest in diameter after three of Jupiter's and Saturn's Titan. The Jovian satellite Ganymede is the largest moon of all.

Over the face of the moon the line between light and darkness progresses at about 10 mph along the lunar equator. Keeping up with it would make for a most interesting moonwalk. From our viewpoint the movement of this terminator is clear to the naked eye from one night to another, as the phases advance then recede.

Commencing with the waxing crescent of a new moon one sees "the old moon in the new moon's arms" in the dusk with faint earthshine illuminating the greater dark portion. Leonardo da Vinci was first to reason this one out and the same effect greets the dawn with a waning crescent phase. Seen from the moon the sun takes nearly two weeks to cross the sky and the earth remains stationary, neither rising nor setting but clearly spinning beneath continents, seas and shifting cloud. From Apollo 8 Borman remarked that it was the only place with any color. We are never visible from the farside and the imposing photos of the world coming out the lunar horizon were taken by orbiting spacecraft.

Strictly, the earth and moon revolve around their common center of gravity, a point lying 2,900 from the center of the earth or about 1,000 miles beneath the ground where it is overhead. The moon's distance ranges from an average 225,622 miles at perigee to 252,088 at apogee along its orbit, a differential of about 26,466 miles in these closest and furthest points. The distance to the moon equals about 30 Earth diameters.

Due to the torque exerted by the sun and on the angular momentum of the earth–moon assembly, the plane of the moon's orbit gradually rotates westward. There are collectively five types of lunar months or defined types of orbit and we present the technical descriptions. Keep up the courage with applied data.

The longest of the five lunar months is the cycle of phases or synodic period of 29.5 days relative to the sun, noticeably longer than its sidereal orbital period requiring 27.3 days relative to the stars. During the sidereal month of the moon's complete circuit of the star sphere the earth-moon assembly travels approximately a twelfth its way around the sun, rendering a lunation cycle just over two days longer. Visually spotting the earliest thin crescent of new moon emerging from the sun is an observational sport. All that calendar stuff would have been easier with twelve equal months in one exact year but nothing was made for our convenience.

Intermediate in length to the synodic and sidereal months comes its anomalistic period, the time between successive perigees. The line of apsides joining perigee and apogee progresses in a prograde direction with a full rotation taking 8.85 years.

In further descending order from the sidereal period there is then the slightly shorter tropical month, which are successive conjunctions with the First Point of Aries. This is linked to the precession of the equinoxes.

Lastly comes the shortest and draconic month, which are successive passages through the ascending nodes of orbit where the path of the moon intersects the ecliptic. A lunar or solar eclipse occurs when the moon actually occupies the ascending node passing N to S or descending node moving S to N with the sun at the same or opposite node.

This is also the 180º alignment, termed syzygy, where spring tides occur the new moon can pass directly in front of the sun or a full moon traverses the shadow of the earth. The two nodes crossing the ecliptic progress in a retrograde direction, and their cycle of 18.6 years, are connectedly equal to the earth's period of nutation.

The positions of the nodes also relates to the cycles and series of eclipses. The moon can pass partially or totally over the disc of the sun to make a solar eclipse and the earth's umbral or penumbral shadows are more than long enough cover the moon at its distance and cause a total or partial lunar eclipse. The umbra or deepest shadow extends over a million miles into space. (Recommended is Duncan Steel's *Eclipse*.)

The moon's orbit lies inclined by 5.14º to the ecliptic and over months, 59% of the moon's face is visible by regular tilting effects that rock the lunar globe slightly back and forth. There are three types of libration. The one in longitude is due to the eccentricity of the moon's orbit where the rotation can either lead or lag behind its position allowing us to view a little more of its western or eastern extremities. It is result of varying distance of perigee to apogee over an orbit whose eccentricity e = 0.0549.)

A second libration in longitude derives from the inclination of 6.7º between the moon's axis of rotation and the normal to the plane of its orbit around the earth where we may peek a little over either pole. Thirdly, there is a daily diurnal libration as a small oscillation that carries the observer by the rotation of the earth.

By sheer coincidence the sun is 400 times larger than the moon and 400 times further away. Hence, they appear the same size, subtending approximately ½º on the celestial sphere. When I was holding forth on the great American eclipse, very few people seemed aware of the statistic or that in true scale, 100 Earths would stretch across the face of the sun.

With very tiny irregularities in its motion we are able to make better measurements than entirely precise predictions of the moon's motions. The gravitational three or multiple body problem has no perfect mathematical solution. Newton said it made his head hurt. Courtesy of the laser reflectors deployed on its surface we can assess its distance to accuracies of inches and the instruments have already accumulated some meteoric dust. Signals dispatched there and back by a 100-inch telescope provided more proofs of general relativity.

Here are some guides to scale and appearances. The moon is 2159 miles in equatorial diameter, a few miles larger than pole-to-pole and covers 14.6 million square miles. It moves at a slightly variable 2,900 miles an hour or 0.8 miles a second at a distance oscillating either side of a ¼ of a million miles. It covers its own angular diameter in about an hour as the world spins and the moon moves on its prograde path about us.

The conspicuous crater named for Tycho, lying in the lower middle, is 54 miles in diameter, and Copernicus to the middle left slightly more. The extensive rays and ejecta blankets from both make them prominent features. Both formed after the most intense period of bombardment and estimates include that Tycho originated a mere 108 Ma ago, long after the event forming Copernicus that occurred some 1.1 Ga in the past. Mare Crisium, the blue and oval shaped patch in the NE, is actually longer across its EW than NS axis as it appears in the top right lunar limb, and its greater diameter is 345 miles across. Perspective gives us a distorted view.

The single most imposing feature, the Orientale Basin is almost entirely unseen from Earth. Only the extremities can be glimpsed at the far SW limb under conditions of favorable libration and illumination. Patrick Moore had suggested it was one wall of a very major feature and was proved correct with the vastly improved imagery. Mare Orientale is an imposing "bullseye" feature surrounded by three concentric rings totaling 560 miles in diameter. Similarly, the largest lava filled crater Tsiolkovsky extends 110 miles but lies on the farside. The nearside crater Plato is another example of this.

The Earth's atmospheric shield effectively protects against continuously incoming tiny meteoroids. Conversely, the face of the moon has gathered meteoric dust over the extreme long term, their inches of depth being an imposing testament to the colossal duration of accumulation.

It is like walking on snow and proved clingy to boots, gloves and spacesuits as astronauts disrobed back in the lunar module. Moon dust was also pungent. In the moment it was remarked to resemble gunpowder, but its composition of silicon dioxides rich in iron, calcium and magnesium bound up with such minerals as olivine and pyroxene suddenly in contact with air probably released the odors. There is a photo of Cernan in the lunar module with serious extraterrestrial dirt on his clothing and face.

The surface is geologically fairly pristine, both in absolute age and comparison to the earth's crust, which is far younger. The moon is some slightly lesser complete age but the first few hundred million years of its existence are as ever, lost to investigation. The assumed molten formation at this primeval stage was probably a global sea of hot magma with any water or associated compounds already lost in hyper heated vapor.

The moon was soon depleted of iron and the volatile elements that form atmospheric gases and water. We noted that the means by which Earth acquired so much water is still uncertain but how the moon immediately lost all of hers is better understood.

There is entirely no weathering, erosion, tectonic/crustal movements or any action involving water even as chemical bonds. These are static conditions in stark contrast to the processes still hard at work here. The estimated

3,000 mini moonquakes a year lie relatively deeper than Earth's but have the collective energy of a firecracker. They are regularly triggered by tidal stains with the earth and occur in distinct areas.

These tiny contemporary moonquakes require surface instruments to register them at all. They were also induced by deliberately crashing expended lunar module ascent stages or third stage Saturn boosters onto the moon plus more modest explosive pops set off just below the surface by the astronauts, all to radiate detectable waves to seismometers. On hearing that the third stage booster of Apollo 13 had hit as planned Lovell remarked "at least something went right on this mission."

Relatively, it is a very dark body. This might sound strange beneath the blaze of a full moon or requiring dark filters at a telescope's eyepiece to manage glare but consider Bond albedo: the fraction of power in the total electromagnetic radiation incident on an astronomical body scattered back out into space.

The albedo for the moon averages 11%, slightly less than Mercury's 12% but much lower than the earth's variable 31%. Mars reflects 25% and the thick atmosphere of Venus a high 90%. The general figure for asteroids is 4% whilst Saturn's moon Enceladus has the highest of all at 99% due to its predominantly icy surface.

Much has been said of the story of lunar exploration and we shall briefly reiterate how emissaries of life from Earth first made the journey. As the former epitome of the unattainable, the whole study of the moon has not been the same since.

Continuing a wide range of flybys, orbiters, hard and soft landers from both superpowers the Russian Zond 5 successfully flew tortoises, flies and plants around the moon and safely back in September 1968.

Two months later Zond 6 also completed another lunar circumnavigation but finished badly due to parachute trouble. One record states dryly that "depressurization lead to biologicals death." Both preceded Apollo 8 and in the height of the space race it was never certain what the Russians were specifically doing. Apollo 11 was in full preparation for launch before Apollo 10 had returned.

In July 1969 the Soviets tried to upstage the Americans in an attempt to bring back some moonrocks before the first manned landing. They launched just three days prior to Apollo 11 and for safety, the Russians actually shared something of the flight plan. In outcome, this Luna 15 lander craft completed 52 orbits but crashed into Mare Crisium when the first moonwalkers were still at Tranquility Base.

In later years, the Russians did manage to return about five ounces of material over three successful unmanned missions and several flops, the first success coming in September 1970 with Luna 16. This was after the second moonwalk of Apollo 12's and the drama of Apollo 13.

I was just a kid in an enthralled global audience, but the high adventure of reaching and walking on the moon impressed people of all ages. It was one of the happiest "where were you at the time?" type moments. Following their return, the crew of Apollo 11 received a sense of shared triumph as they took on international tours, basking in their well-earned glory.

By the aegis of applied science, the moonrocks supplied answers but evoked more questions. Geological samples were a specific objective as was the deployment of the Apollo Lunar Surface Experiment Package. Conclusively, the darker highland and rugged areas are older than the maria regions and the 843 pounds of samples returned from six different locations proved entirely igneous having solidified from fiery volcanic processes. I gaze in wonder at the sample in Seattle's Museum of Flight. There is a photo of it at its place of collection at the astronaut's foot.

In simplified conclusion, the oldest surviving rocks on Earth are approximately equal in age to the moon's youngest. The so called "Genesis rock" collected by Apollo 15 is dated 4.1 +/- 0.01 Ga in formation, a mere 100 Ma less old than the solar system. The most senior terrestrial rocks are an estimated 4.28 Ga in age and the lunar highland samples range from 4–5 Ga. In comparison to Earth, parts of the Canadian Shield are reliably dated at 4.031 +/- 0.003 Ga with others of similar pedigree in Africa and Australia as 4.28 Ga in age.

There is no tectonism, active volcanism or surface movement and another reason the moonrocks are so well preserved is that there was no reaction with water to form any sedimentary materials such as clay, shale or limestone. Any reactions were entirely chemically anhydrous. In 2013 new techniques applied to Apollo 17 samples revealed some small amounts of rare volcanic glasses containing water at a mere 30 parts per million and their source was some fire fountain active Ga ago.

Any quantities of water ice that might exist at the permanently shadowed poles probably originated from comets and water-enriched asteroids. There are no bodies of water on the moon and no hydrological or other cycles but it is often said that such potential reserves could prove of use to future lunar bases.

As for temperatures, they can vary from 123º C to −173º C over a typical day with record lows of -247º C and −238º C. These were recorded at the north and south polar regions respectively where there are cratered regions unpenetrated by any sunlight or heating. As recorded by the Lunar Reconnaissance Orbiter still in orbit these areas are only 26º C from absolute zero. Note these temperatures to be some 7º C colder than Pluto, the previously assumed place of maximum deepfreeze and still a bigger chill than the atmospheres of Uranus and Neptune.

Similar conditions prevail on Mercury where diurnal temperatures plunge from 427º C to −170º C. We do not have to venture far to find the coldest natural place in the solar system. We can directly see it.

There are an estimated 181,000 craters on the moon larger than 1 km. Before we understood the Late Heavy Bombardment in context and causation it was termed the Lunar Cataclysm. The processes took place 3.8–4.1 Ga ago and probably later in the formative accretion period when the terrestrial planets were finalizing their masses. One contemporary meta theory invokes the gas giant planets shifting outward in their orbits at this stage and directing vast quantities of material inward. The moon and Mercury bear the scars to this day. The Nice model (named after the city in France) is one scenario of the dynamical evolution of the solar system with such a period of bombardment of the inner worlds by asteroids. There is abundant surviving evidence.

This links to questions like the age of the enduring asteroid belt as a collective zone compared to its individual components. Did they entirely form where they are or did they gravitationally congregate between Mars and Jupiter? The former is more likely. Asteroids are primeval bodies and Jupiter's gravity is highly influential, Trojan groups and Kirkwood gaps included. Overall, the Nice model differs from other hypotheses in proposing orbital expansions long after the dissipation of the initial protoplanetary disc. It does meet limitations in easily explaining the satellites of the outer solar system and the Kuiper belt. Another school of thought has Uranus and Neptune forming more slowly than the others.

The same episodic impact processes left their dramatic marks on Mercury, as derived from crater counting techniques on this other mostly non-reworked surface. There too, the size and regularity of impacts dropped off after an intense period. Mercury lacks the features of lava plains but much of its face otherwise resembles the moon. The Caloris Basin on Mercury is *possibly* equivalent in age to the younger lunar maria. Other planetary satellites show a huge range of cratering and/or subsequent activity causing resurfacing whilst Earth and Mars have long since erased any direct evidence of the LHB.

When the Voyagers were at Jupiter and its principal moons were properly revealed I felt that Callisto was the only member proving entirely as anticipated; a dark, unchanged and crater dominated wilderness.

The relative smoothness of the maria indicates how the flows occurred sometime after the period of greatest bombardment. They have not been around as long as the heavily pockmarked plains and highlands. Later, the big basins carved out by giant impacts were filled by massive flows of lava to form the maria. This occurred 3–3.5 Ga ago and suggests the existing basins to be some 500 Ma younger in process. We deduce that there is no direct causal relationship between the impact events and the mare volcanism that made the fills.

Our model of lunar evolution proposes that heat produced by the decay of radioactive elements subsequently melted the lunar material at depths of some 125 miles. Only later did this cause a great flood of lava to emerge, possibly in stages taking that half a Ga in themselves. Curiously, the apparently youngest flow and largest expanse forming Oceanus Procellarum does not correspond to any known impact basin.

In context, the early history of the moon was quite turbulent but long ago became geologically quiescent. A very little outgassing does still take place as TLPs (transient lunar phenomena) as observed in specific areas, notably the crater Aristarchus.

Mascons ("mass concentrations") are a squashing of crustal material by both internal and impact forces to higher densities. In causing slightly higher localized gravitation, their influence is detectable as tiny varying pulls on satellites in lunar orbit. This is useful for detailed gravitational mapping, well accommodated by NASA's Grail mission.

The surface termed the regolith is a layer of broken up powder and rubble about 3–60 feet deep consisting of loose heterogeneous material covering the solid rock. From the Greek *rhegos* for blanket and *lithos* for rock, this covers almost the entire lunar surface with bedrock only protruding on very steep sided craters and occasional lava channels. We once feared that a descending probe would disappear in beds of deep dust and surface exploration was a non-starter.

This broken up top layer or lunar "soil" is quite unlike the equivalent in our gardens having been built up by the continuous bombardment of meteors and cosmic dust but the process is called "gardening." Past impacts served to shatter the solid rock, scattering material by stirring and mixing the lunar soil. The Apollo samples include fragments whose collection sites lay far from their sources.

The lunar crust averages 37 miles thick and is slightly thicker on the central farside by about 7½ miles. This leads down to a mantle extending over 500 miles further in depth. We can deduce that the core is some 300 miles in diameter, relatively small with an assumed molten outer and solid inner structure. It is definitely slightly offset in the direction of the earth, another sign of their long-term association. We do not know its exact chemistry or temperature but estimate some 1300 ºC. Certainly no heat reaches the crust or surface.

We can provide better ideas for what the surface is made of. Elements present include oxygen, iron and silicon in greatest abundance followed by magnesium, calcium, aluminum, manganese and titanium and other minerals. Carbon and nitrogen occur only in trace quantities deposited by the solar wind. When Apollo 8 returned, Anders proudly remarked that it was made of American cheese.

Let us tackle one specific mystery before studying the prime question of the origin of the moon. Three current possibilities exist to explain how the uppermost layers are thicker on the farside than the nearside. The prime question here is whether the causes were bound up in the lunar genesis or by effects playing out later.

Most likely, the huge impact forming the South Pole Aitken Basin may have internally shifted significant amounts of material away from the point where it hit, shifting internal material. The deepest crust lies antipodal to the approximate center of impact and this single event was almost violent enough to have shattered the moon. It must have happened early in its independent biography.

Lying eight miles beneath the average level this basin is the lowest elevation in the lunar crust and the largest generalized crater formation anywhere in the solar system. Named for two features at its extremities, its 1600-mile diameter is so large as not to be obvious as a basin. (For mountaineers of the future, the slow gradient of Olympus Mons on Mars would make it hard to appreciate that one was climbing the biggest mountain in the solar system.)

Whilst greater crustal thickness possibly restricted the quantities of magma able to reach the surface the theory cannot easily explain how the resulting thin crust of the South Pole Aitken basin itself did not become more fully filled by the products of lava. It is not a mare as such.

A second possibility is that the slightly differing depths formed under the full influence of tidal interaction with the earth. It proposes that the crust facing this way was formatively restricted to thinner concentration and assumes purely developmental causes in the solidification process. Whilst the lunar crust floated on an ocean of molten rock and the young moon was much closer, tidal effects created distortions that became frozen in place. The results endure.

It also requires that the captured rotation of the moon settled down extremely early, probably too soon for the theory to work successfully. Dynamically, the centrifugal force of the moon's rotation balances the earth's gravitational attraction and the spin of the primeval satellite probably did not retard and stabilize so very soon.

A third scenario to explain the thicker farside crust goes thus: when the actions of the LHB formed the sites for the future maria on the nearside the impacts showered molten rock and deposited great sheets of magma on the other hemisphere. The smashing of big asteroids in very short order on only one side of the young moon could simply explain the difference of the two hemispheres

Some combination of these factors may have done the trick including asymmetric nearside/farside cratering. Let us apply Occam's razor and assume the simplest explanation. Lava outflows were certainly prevented and did not form any significant outflow on the farside with the exception of Mare Moscoviense.

We could additionally propose a second proto satellite or planetoids crashing moonward to significantly add to the regionalized thickness of crust. We have no end of asteroids to hurl around and plenty of past time for such conjectures.

Now for the origin of the moon, a question still steeped in mystery. Despite enormous study we still do not unequivocally know. We can regard Earth and moon as bodies of interrelated origins with our satellite of some slightly lesser age. So how to make the moon?

After much debate, we opt for the theory of Collisional Ejection or Giant Impact, also inelegantly known as the Big Whack or Big Splat (ugh) as the working hypothesis.

Collisional Ejection is characterized by a major collision of some body with the earth, ejecting huge material that went on to form the moon in orbit. Captured rotation takes time and we appreciate that the deep adaptations of marine and other life to tides to similarly reveal extended timeframes. By various lines of reasoning we've been together a long time.

Following the giant collision, we posit most of the ejected mass incorporating relatively quickly into a single body. There are certainly no other chunks left over in any Earth orbit now and we have no "other moons." (There are cases that some known tiny asteroids are very temporarily engaged by our gravitational influence. Review 3753 Cruithne, a minor asteroid with 1:1 orbital resonance with the earth in a horseshoe orbit or 2006 RH$_{120}$ which was originally mistaken for a piece of space junk.)

Can we make Collisional Ejection work?

Propose the major fragmentation from a Mars sized body (?) to have collided with the primeval Earth. Sending out a huge cascade of material that aggregated into the embryo moon. We propose that the components aggregated into a single mass in a period of perhaps mere centuries and took up orbital residence at perhaps one tenth of the moon's current distance. Unfortunately, all the associated time frames are difficult to assess despite the best attempt computer simulations.

Amidst so much speculation of formation, let us ask how matters ended up. In mass, the earth was rendered 81 times that of the moon, 4 times larger in diameter and a factor of 50 greater in volume. The Earth has a greater higher general density and the surface gravity is one sixth of a terrestrial g, which proves convenient in hefty spacesuits and lifting loads. Collisional Ejection is the least bad of the bunch but it comports an unavoidable vagueness and overall, we had hoped for better by now. It also infers that the very young Earth was alone.

The Earth has a large iron core, unlike the moon because the iron presumably had sunk to the center of the earth before the impact. The existing iron core of the moon probably takes up some 25% of its radius compared to the earth's 50%. Its relative smallness is further evidenced by the very low external lunar magnetic field. There magnetizations now present are probably crustal and the dynamo within the earth produces a far greater field. Therefore, was it the iron depleted rocky mantles supplying the material that went to form the moon with the iron core of the impacter melting to merge with that of the earth's? That the moon generally lacks iron is consistent with its average density of 3.3 gms/cc compared to the earth's 5.5 gms/cc.

The other competing theories of lunar origin included a Fission model, Intact Capture and another hypothesis based on Binary Accretion. These alternatives are all generally written off by now but they are entitled to a hearing.

Unfortunately for the Fission model there are serious dynamical objections. Consider the unequivocal faster rate of rotation of the earth in the great past as the mechanism of separation to create the moon. A single piece of such mass spun off by the spin of the earth and breaking free from its gravity would have possessed enough momentum to have entirely headed off into space. It would be long departed as the fifth terrestrial planet or biggest asteroid. The Pacific Ocean basin is far too young to be any scar of its departure, as once suggested. It is coincidental that their diameters roughly match and not viable that the earth shed a significant piece of itself from any site that can still be seen. Studies of matched isotopes suggest that fission was partly the case but this is more fully accommodated by Collisional Ejection.

Now we rule out the Intact Capture hypothesis. At a current distance of thirty Earth diameters and increasing, the earth and moon is like an established double planet. The condition that they were definitively closer in the great past immediately weakens any case for a moon spiraling closer into stable orbit from some planetary distance. In short, our neighbor could not have formed indèpendently and then joined us. It is quite untenable that this took place before commencing to move slowly away again. Here is a hypothesis directly slayed by a fact. Finally, a close approach by bodies of their masses and assumed speeds at this distance from the sun would more likely have resulted in either collision or radically altered trajectories. Not an orbital merger finalizing with an eccentricity as low as 0.0549.

As for Binary Accretion, we found major geochemical objections to any proposed original development out of the same primordial disc. Earth and moon did not form together as separate bodies from the same protoplanetary material, the accepted case for the greater solar system. It also cannot expound their different periods of rotation. Jupiter and Saturn may well have given rise to their larger moons by their own processes of binary accretion. The satellites still orbiting in their equatorial planes probably did originate this way and captured rotation is further evidence. There was certainly the mass to spare with these solar systems in miniature. Intact captures of previously independent asteroids probably applies to the tiny Martian moons Deimos and Phobos. It has been said that satellites are fossils of the primary's past.

If Collisional Ejection becomes the prime hypothesis, we necessarily ask what that other colliding body was. It even has a ready-to-go name with "Theia" as the mythological mother of the moon goddess Selene. In suggesting that Mercury played the crucial role we may stretch the concept to explain how the smallest planet has such an anomalously large core. It lost all its original crust and some part of the mantle in the titanic glancing impact.

But it goes too far to invoke Mercury. The evolution of Mercury's spin-orbit resonance preludes it ever crossing the path of the earth. It has a notably elliptical orbit but its path never reached out this far. The circularity of the orbits of both Venus and Earth strongly suggest that they are long estab-lished at present distances.

Was Theia partly absorbed into the earth *and* the source of material con-tributing to the formation of the moon? We could radically tweak the idea by invoking Venus as the colliding body, explaining its own slow and retro-grade spin as a result of striking Earth a glancing blow. Again, this is pure speculation. Accepted and disappointingly vague as it is, the model of Colli-sional Ejection should not be pushed to expound *too* much. We risk sound-ing like Velikovsky, who was considered the wicked witch of the woods of cosmogonic hypotheses when I began astronomy.

Sadly, we may never know the truth. It all took place too far in the past to be elucidated by present science and the trail is as cold as the lunar poles. Theia is of posthumous significance and the moon still guards several of her secrets.

Let us retrace some reliable conclusions. The visible moon endures most-ly unchanged since the LHB, the later maria eruptions and a dwindling pro-cess of cratering. Extremely minor such events still occur and the tiny flashes of contemporary meteor impacts have been successfully captured on film on occasions.

By all accounts, the moon has held essentially the same for > 90% of its existence. For two worlds so closely locked in a mutual gravitational field for over 4 Ga there are marked differences between them. Here on Earth there is ceaseless geophysical (and biological) activity and there on the moon there has been none in general for 3.8 Ga.

Their paths of planetary evolution soon diverged, most significantly the acquisition or lack of reducing atmospheres or the absence or abundance of water.

The complete dearth of biological activity allowed us to confidently abandon quarantine for returning astronauts after the third landing. There are no microorganisms on the moon but it was best to play it safe at first. Journeys into the unknown could pack nasty and unforeseen surprises.

The tiny Earth bugs in the camera of Surveyor 3, whose pieces were col-lected and brought back by Apollo 12's astronauts proved at length to be laboratory contaminated. Back in the receiving facilities studying moonrock this first created quite a stir but conclusively, the organisms had *not* been accidentally flown to the moon in the first place and there was no question of surviving 2½ years in the sterility of the lunar environment and brought home intact. We meet a certain redivivus that space remains a full natural quarantine.

It is possible for armchair astronauts to visually share the experience of being on the moon. Look closely at the film footage of the landscape hosting the moonwalks and moondrives and see how the distances of background features are difficult to assess. That the horizon seems deceptively close is partly due to directly being a smaller globe but features seem anomalously close. It is not easy to gauge the distances of background hills and craters and they are certainly further than they appear. It is what our brains are used to seeing and telling us. Also, shadows are stark and there is no twilight or softening of the hues of sunlight by air.

For the safety of manageable temperatures, all our EVA's were timed for the lunar early morning. The glare of day is too bright to see much of the bright point sources in the sky other than the bolder, unfiltered sun slower across the sky. Night in moon orbit beyond the light of both the earth and sun must be truly awe-inspiring. I once had the pleasure of meeting Fred Haise of Apollo 13 and listened enthralled to a first-hand account of Tsiolkovsky crater.

As for the earth, our jewel of a cosmic oasis never looks the same twice in its unsurpassed beauty as the great cosmic oasis. Several space travelers shared the view that everything they ever knew and the sum total of human cacophony takes place right there. Against the eternal backdrop of stars it does bob up and down a little at what we call the libration zones but no human eyes have yet seen a lunar eclipse from the moon. Imagine the earth, four times larger than the sun eclipsing it. As for going through phases equal and opposite to the moon as seen from home, is this the most evocative demonstration of Newton's third law?

My late father once mused how sad it was that the moon was not another mini Earth in full and abundant attributes. Ideally, this would be a ready-made new world complete with forests, rivers and resources akin to our own with the added non-attraction of lower gravity. If only the moon duplicated the lush glory of Mother Earth in our first stepping-stone to the stars. Assuming we could live there directly we might have been inspired to reach it earlier.

It would have been a busier dream for human endeavor than Lucian, Kepler or Cyrano De Bergerac suggested in their literary works, the earliest imagined voyage being written in 79 AD. Realistically, Verne's yarns were singularly influential for how we might get there as a feat of engineering. As the vehicle of pragmatic choice, the development of big rockets first built as weapons might have taken a more genteel course.

Visualize cosmic nature extending the invitation of a full-scale binary planet system. With forests and real seas on show, we would have perceived the moon very differently. The proximity of a whole Earth Junior would have affected our entire worldview but instead the moon has always been lifeless and inhospitable. Despite the enthusiasm of generations of space dreamers the immediate impressions of the first humans to go there included "forbidding" and "a distraught kind of place."

That the moon is decisively not another little Earth in waiting is the most colossal of lost alternate futures for Mankind. How different our spacefaring future would be with such a convenient next-door neighbor for a first step to the stars. Think what could be accomplished without any necessity of life support. We went all the way to the moon yet it was the sight of the earth in space that we truly discovered for the first time.

Primordial Soup Of The Day

We must try not be overwhelmed by the unimaginable timespans inherent in Earth history, both geophysical and the long term biological. Such scales defy comprehension and it is the same with the distances we encounter raising our eyes to the night sky. We must train our scientific thinking to the grand vistas of both size and proportion, sparing a sympathetic thought for Pascal whom reputedly became frightened by its scales. The vistas of time and space indicate some meaning to our existence or at least a sense of place.

We must make the attempt in the abiding spirit of science. For example, we rationally place ourselves in the Holocene epoch of the Quaternary period within the Cenozoic era all commencing that 65 Ma ago when that crucial page was turned.

The ongoing Phanerozoic Eon reaches back to the dawn of the Cambrian period and was preceded by the Precambrian Eon, which extends downward to include all time prior to 543 Ma past. We have constructed some meaningful depth lines into Earth history but admittedly, it is only the shallower ones lending themselves to direct study.

One of John Locke's analogies comes to mind. It is of great use to the sailor to know the length of his line though he cannot with it fathom all the depths of the ocean.

Moving up to date the Greek term *holos* means whole and *kainos*, new. Holocene therefore describes the entirely current and "wholly new" epoch commencing with the warming characterized by the retreat of the last great ice sheets. We are presumably in some interglacial period at present because they periodically occur over different periods and severity.

About ice we encounter the anomalous contraction of water and the curious fact that water in its solid form is less dense than its own liquid, freezing point being the defined 0° C on the scale of temperature. Ice floats on water whilst most compounds sink in their own fluids or melts. Whilst the sights of icy waterways or icebergs afloat are common, this is actually a very curious effect and played a crucial part in the story of life.

It means that ice can effectively provide a protective surface layer over large bodies of water. Indicating that some marine life continued unimpaired throughout long glaciation events. Beneath the solid ice cover conditions can stay adequately warm or at least at stable temperatures indefinitely.

Lakes and seas freeze from the top down and no such benign protection was present for the landmasses as the huge glaciers came and went. Of late, we are learning much about life in the Antarctic, where much more biological action is going on than previously supposed. Extremely hardy and patient organisms exist there.

Throughout the long biological career of Mother Earth things have changed at a range of paces. This applies to both the greater environment and the progress of life. The Gaia Hypothesis is a rational description should we be disposed to seeing the totality as a single and self-regulating complex system where life dominates. The concept of one vast organism might be a stretch but remember the SF tale *Solaris*?

The fossil record indicates major setbacks, bursts of activity and many periods of consolidation. There was nothing but bacteria for perhaps 3 Ga from the outset and it seems likely much of this was well below the surface of the highly active crust. Defining biomass as the quantity of organically bound carbon, bacteria are still in excess of plants and animals so in a way, the multitudes of tiny bugs still rule the world. The whole position endures so welcome to the planet of the unseen bacteria and microorganisms. For all they do in global ecology it is a real contrast to lumbering lizards the size of buildings.

As for extrasolar planets orbiting other stars, they were previously extremely difficult to detect and could only be done indirectly and uncertainly. It was only the dynamical effects revealed in planets' orbital interrelations with parent stars and fortuitous transit events that suggested their presence.

It took a lot of work to forge any progress and there were a few false starts to establish any details until the Kepler mission changed everything. Observing from above the earth's atmosphere has distinct advantages and the data are still in the process. There is now every reason to suppose that exoplanets are common to the extent of making the provisional statement that most stars possess them. It presents a position long thought to have been the case that there are plenty of accommodating stages with the potential for life.

How did biogenesis take place? Here is a précis of five possibilities with the first two preferred as descriptions. Perhaps we should not restrict the speculation to a single means taking place in one locale at merely one point in time. The origin of life could have been a combination in different places and times in the primeval depths of Earth history.

The Primordial Soup Theory

In the 1920s Haldane and Oparin proposed organic chemicals accumulating on the surfaces of bodies of water beneath the hydrogen rich younger atmosphere before the release of free oxygen. The early Earth had a chemically reducing atmosphere that by the exposure to several forms of energy produced simple organic compounds known as monomers.

These compounds accumulated in a "soup" that became concentrated in several locations like shorelines and oceanic vents, transforming into far more complex organic polymers. It is unknown how such polymers progressed to protocells. It is problematic that in the aqueous environment their hydrolysis would more likely break them down into constituent monomers rather than condense them into polymers.

Deep Sea Vents

In the depths of the oceans hydrothermal vents provided the templates of the first true cells in the thin mineral walls of the interconnected vent micropores. It was the action of primitive enzymes for the reduction of carbon dioxide with hydrogen to the formation of organic molecules. The hydrogen was produced by the contact of minerals contained in seabed rock with water.

Such hydrothermal vents or black smokers in the marine environment are still active as fissures commonly found near volcanically active places. Relative to the majority of the deep sea the submarine vents are generally more biologically productive. In all scenarios, replicating RNA was a prime step, leading to the debate of which came first, metabolism or information? The important question is how closely do geochemistry and biochemistry align?

Panspermia

Here is the suggestion that life is not entirely indigenous but was fully imported as tiny organisms or prebiotic material arriving on asteroids and comets. It evokes the scenario of microscopic life being brought to the hospitable conditions of the early Earth and is therefore more a mechanism of distribution from original sources. Organic molecules certainly exist in interstellar dust and panspermia proposes that this medium is sufficiently protective from radiation for indefinite periods. Pre biotic molecules could have developed in the original proto planetary nebula.

Since then, their propagation is possible on any scheme of comets, radiation pressure or microorganisms embedded in rocks that finally arrived on Earth. Pushing the hypothesis to extremes, Earth was accidentally seeded by rocks from Mars or deliberately engendered by the action of aliens or a supernatural agency.

Deep Hot Biosphere Theory

Below the earth's crust there was a continuous supply of abiotic hydro-carbons, principally primordial methane within the mantle driven by chem-ical sources of energy. Solar energy and photosynthesis were not involved in the speculation that prokaryotic (the unicellular organisms without mem-branes or organelles) massively developed in the mantle, slowly building the means to conquer the surface or sea bottoms by sheer biomass of slow up-ward movement. Back then the atmospheric conditions were not yet crucial-ly relevant. Rather the earth's inner heat sources of trapped primordial heat and radioactivity are held as adequate.

Perhaps heat loving bacteria existed in such profusion to have achieved this. The hypothesis is largely the proposal of Thomas Gold and the question for the future is how biofriendly are the subsurfaces of planets?

Radioactive Beach Hypothesis

This posits the collection of radioactive material by powerful sorting tides generating the complex molecules leading up to carbon-based self-rep-lication. Radiation might seem antithetical to biogenesis as it tends to break chemical bonds and split larger molecules but it can act to provide the chem-ical energy required to generate some of the basic building blocks.

An obvious objection is the lack of atmospheric oxygen preventing ura-nium becoming water-soluble but much stronger tidal processes could have concentrated radioactive grains of uranium on some primordial beach where they generated proteins. It was therefore radioactivity that set off the first construction of independent cells. Under any scenario of biogenesis it took place at a time when the moon was closer and the earth span faster.

From the perspective of life unequivocally commencing relatively early on Mother Earth, it throws up the following powerful questions:

1. Is the universe biofriendly?

2. If life began so quickly on the primeval Earth does this equate to bio-genesis generally starting soon on young worlds? The curse of the single ex-ample dogs all our efforts.

3. More precisely, if conditions are right do biochemical reactions immediately and busily ensue to eventually bring a new planetary biosphere into being?

Obviously, it only occurred on Earth (among the immediate planets) and as chemical processes of ever-growing complexity and diversity, it is anything but the decline into disorder that characterizes entropy. It looks a long road ahead for geophysics, chemistry and biology to expound and successfully inter-relate the causation of biogenesis. What can be said of this?

Life crucially requires the interaction of carbon and water and all organisms need four types of organic molecules: Nucleic acids (DNA and RNA), proteins, carbohydrates and lipids. Genes are made up of DNA and each gene provides the code for making a specific protein. They have been called the very blueprint of life. DNA bears a twisted double helix and RNA assumes many different shapes.

We observe that proteins are the most versatile, made up of about twenty different amino acids in combination. Carbohydrates contribute the greatest quantities of organic molecules and are basically composed of sugars accompanied by lipids containing the most energy.

Who else but Darwin shall we turn to? He once stated that the origin of life would be as elusive as the origin of matter to establish. Happily, there has been great strides since his day and it is as important to ask the right questions...or at least pass on the genes of mature curiosity as an advantage in natural selection favoring our own species. Before we frame a reply to him knowing what we do now we will tackle the frontiers of cosmology. We endeavor to ascertain the grand design of the cosmos.

CHAPTER 3. THE NATURE OF THE UNIVERSE

Fiat Lux

The electromagnetic spectrum comprises the total range of radiation made up of the waves of energy emitted by charged particles. It is a vast family of hugely diverse wavelengths and frequencies of electromagnetic energy whose smallest units or quanta are regarded as massless photons.

What may be directly seen of it? Visible light proves to be a very narrow part of the full spectrum. Immediately beyond the limit of human vision at the red of the visible spectrum lies radiant heat, which we can feel if not see in the direction of longer wavelengths and shortly frequencies. At the other end of this visible band beyond violet light come the shorter wavelengths and higher frequencies of the ultra violet. Most of our empirical knowledge derives from the detection and interpretation of these electromagnetic waves by both simple and sophisticated optical and radio equipment.

In the everyday world glass and sprays of water are often seen dispersing sunlight, acting as natural prisms to break white sunlight into its components of colors. This was first established by Newton in an experiment in a dark room allowing a narrow beam of light to enter via a slit in the black shade. When refracted by the prism each color is bent by a different amount and its emergent beam diverges into a spectrum. In definition, the dispersion of light is the splitting of white light into its constituent colors due to the refractive index of the surface and the different wavelengths of the light. As for rainbows in the sky, Spinoza expounded that the same action allows water droplets in the atmosphere after rain showers.

As perceived by the human eye, visible light ranges from wavelengths λ (lower case Greek letter lambda) and extends a mere 390–790 nm from crest to crest. A nanometre is 10^{-9} of a meter and atoms range from 0.1 to 0.5 nm in diameter. For frequency f light is equivalent to the range of 430–790 THz where one hertz is a cycle per second and a terahertz is a multiple of 10^{12} Hz. Wavelengths are inversely proportional to frequency with more energy packed into the shorter frequencies.

So one of the home truths of human perception of the external and phe-nomenological world is that we directly sense a miniscule part of the full electromagnetic spectrum. The analogy is a piano playing in full swing but we can only hear one note. Above our heads, the earth's atmosphere has a certain optical window and the blue light of shorter wavelengths from the sun is more scattered than the longer and redder, creating the daytime sky. On Mars it is pinker. The envelope of gases surrounding the world is protec-tive in several vital ways of holding out dangerous radiation from space as does the global magnetic field.

Until the age of science little was understood of the earth's magnetic field including its association with the behavior of compass needles and the color-ful displays of aurorae. As usual, superstitions attached to such phenomena *e.g.* the appearance of St Elmo's fire was taken as a good omen for ships at sea rather than electrical discharges during thunderstorms creating luminous plasma at the tips of masts.

About aurorae, Cavendish's triangulations first showed the approximate altitudes of the Northern Lights and Birkeland's experiments at the turn of the 20th century linked them with excited gases high in the atmosphere. These aesthetic displays are incidental to the sterling services provided by the magnetosphere. It acts to deflect most of the solar wind whose charged particles would otherwise strip away the layer of ozone in the atmosphere and allow other damage by their penetration.

High ultraviolet, X rays and gamma rays are termed ionizing radiation because photons of high frequency possess the energy to ionize molecules and break down chemical bonds. They may cause chemical reactions to the detriment of living cells beyond the simple heating of lower frequency waves.

DNA may be directly or indirectly damaged by ultraviolet light to the extent of skin damage and cancer.

As for how we see, the medical theory of vision posits that the change in bonding of a single molecule in the human eye absorbs light in the rhodopsin in the retina.

The electromagnetic spectrum consists of oscillating electric and magnetic fields moving perpendicular to the direction of their energy and propagation. The acceleration of charged particles promotes these electric and magnetic fields to vibrate transversely and sinusoidally at right angles to each other and to the direction of motion. They are permanently in transit at c with no gaps in the continuum. (Spectral lines of absorption and emission result from the conditions in stellar atmospheres.)

Historically, John Herschel first discovered infrared radiation and Faraday linked electromagnetic radiation to electromagnetism in 1845. Electromagnetism is the interaction of electric currents or fields and magnetic fields. Fifteen years later Maxwell derived four partial differential equations for the electromagnetic field, two of which indicated the behavior of waves propagating at the (roughly) known speed of light. It suggested a huge spectrum to exist, now classified into eight classes from the shortest wave gamma radiation through X ray, ultraviolet, visible, infrared, terahertz, microwave and radio waves.

Every body radiates thermal energy but molecular motion ceases and entropy reaches its minimum value at 0 K (for Kelvin) equivalent to -273.15º C. This is suitably termed absolute zero on this temperature scale where there is no colder. The Rankine scale places 0 R at the same minimum point but Rankine degrees count through the Fahrenheit scale commencing at an equal -459.67 º F for absolute zero. According to the laws of thermodynamics 0 K cannot be reached by thermodynamic means. We've heard of "keeping your cool," but why is this so?

The temperature of the substance being cooled approaches the temperature of the cooling agent asymptotically i.e., with a line or curve that approaches a given curve arbitrarily closely. No adiabatic process commenced with a nonzero temperature can lead to zero temperature and no procedures could reduce a system to 0 K in a finite number of operations.

Laboratory research have set records of approach in reaching down within a billionth of a degree of absolute zero by cooling the nuclear spin of a piece of rhodium metal as low as 100 picokelvins. (This unit pK equals 10^{-12} of a Kelvin.)

However, other experiments close to 0 K shows that where molecular motion ceases quantum effects such as superconductivity, superfluidity and Bose-Einstein condensation occur. A system at absolute zero still possesses quantum mechanical zero point energy, the energy of its ground state. The kinetic energy of the ground state cannot be removed. Like c by acceleration, the level of 0 K can never quite be reached.

Where are the hottest and coldest known places? We shall probe much farther afield than California's Death Valley or the Antarctic winter, which are the greatest terrestrial highs and lows. A new high record was set for the former in summer 2017.

The lowest observed natural temperature for a specific place is within NGC 2440; the Boomerang or Bow Tie Nebula located about 5,000 light years away in Centaurus. There, the outflow of gaseous material has dropped conditions to 1 K, making this proto planetary nebula the only known locale with a temperature lower than the CMB. It is the coolest known place in the universe apart from some cold spots within the microwave background.

The extremely low thermal state of NGC 2440 is probably due to a rapid expansion, where some 1,500 years of activity has rendered the nebula about 1 light year across. The estimated present speed of gaseous escape is 164 kms/sec. Millimeter size dust grains mask sections of its center so most escaping visible light forms an unusual hourglass in general appearance. A larger spherical body of cold gas seen only in sub millimeter radio wavelengths surrounds a distinct dual lobe structure. NGC 2440 will likely dissipate in 30,000 years to reveal a single white dwarf star smaller than Earth.

In the generalized case of the background radiation the photons have been shifted into microwaves to a temperature of 2.725 +/- 0.001 K. As the CMB originated from a spherical surface it is known as the surface of last scattering.

To the opposite extreme there are no known upper limits to temperatures. The hearts of supernovae as the active centers of such explosively active stars are the commonest individual cases where conditions allowing up to 100 billion K can occur.

With supernovae, the reserves of nuclear fuel normally lasting for billions of years are being consumed in mere weeks of activity. As for the highest energy output of all, gamma ray bursts and their afterglows of longer wavelengths have been observed from extremely energetic explosions in distant galaxies lasting between milliseconds or minutes.

The causes of GRBs include supernovae or hypernovae or a rapidly rotating high mass star collapsing to form a neutron star, quark star or black hole. Colliding neutron stars could hypothetically elicit temperatures of 350 billion K and whatever their sources, GRBs they are the hottest discrete electromagnetic events in the universe. In general more general terms, the early stages of the Big Bang involved temperatures "dropping" to 100 billion K after one hundredth of a second.

A little more background is helpful on matter, waves and energy. Try and grasp some technicalities described in the following brief primer on applied physics but don't dwell on the mathematics if it's cumbersome.

The longest measured wavelengths and lowest frequencies are radio waves. Their crests and troughs can cover thousands of kilometers. Ranging downward to the very shortest λ and highest f are gamma and cosmic rays with wavelengths in the region of 10^{-16} of a meter, much smaller than atoms.

For the entire electromagnetic spectrum, velocity v is the product of λf. This provides the useful tool:

$$v = \lambda f$$

In the vacuum of space, c replaces v.

Velocity v is reduced if the waves travel through a given medium of refractive index n where

$$v = c/n$$

Emerging from such a medium as water, light immediately regains its maximum velocity.

When the behavior of light is investigated by simple slit experiments, it is not long before curious sights emerge. Whilst the reflection, refraction and interference of light may be described in terms of wave motions, other matters including the photoelectric effect indicate better explanations in terms of particles.

Whether light was a corpuscular particle or wave was another long lasting controversy. Classical physics is unable to fully describe the behavior of quantum scale objects and light can be described in both terms of waves and particles. These are the bases of particle-wave duality and quantum theory where electromagnetic radiation is perceived as a stream of quanta.

Newton had mostly favored a corpuscular version of a theory of light but equivocated now and then. In the same era Huygens tended to a more wave-like concept. Later and more sophisticated experiments show that particles have a wave character and *vice versa*, leading Bohr to see the "duality paradox" as a fundamental metaphysical fact of nature. When Heisenberg replaced the classical field values of Maxwell with quantum field theory determinism was banished for evermore. The snappiest expression of the duality of light is that it propagates as a wave but interacts as a particle. As Einstein wrote:

> It seems as though we must use sometimes the one theory and sometimes the other, while at times we may use either. We are faced with a new kind of difficulty. We have two contradictory pictures of reality; separately neither of them fully explains the phenomena of light but together they do.

Note that he was awarded the Nobel Prize for his "services to theoretical physics and especially for his discovery of the photoelectric effect." It was not actually for the Theory of Relativity. The crux of both is that:

The energy E of the quanta is directly proportional to the radiation's frequency as:

$E = hf$ already known as the Planck relation

where h is the Planck constant equal to $6.62606597 \times 10^{-34} \, m^2 kg/s$.

And most famously amidst the equations of relativity:

$E = mc^2$

Known as the principle of mass-energy equivalence. It shows that energy and mass are interchangeable with each other and that the mass (m) of an object or system is a measure of its energy content E. Enormous energy is therefore available in efficient nuclear processes acting to convert mass to energy.

Playing with the figures for $E = mc^2$ reveals that a single kilogram of mass could fully be converted into 25 billion kilowatt hours of energy or an equivalent 695 million US gallons of automotive gasoline. A pertinent comment is that fossil fuels as a source of energy are notoriously inefficient, let alone non-renewable and pollutive.

On fully extended principles, the hypothetical scales of wavelengths in the electromagnetic spectrum could range upward to the size of the universe itself and downward to the vicinity of the Planck length. This is however taking things too far. It is akin to stating that every body above 0 K radiates across the entire spectrum or that a ray of light circumnavigates the curved spacetime continuum in one eternal moment. Such momentous conclusions are valid but science does not function by stating extreme sets of conditions then filling in the blanks. Let us turn to what happened to create the universe and what we can see of it.

Big Bang And The Observable Universe

We commence with the conditions that apparently built the whole matrix of space and time. From the phenomenally energetic outset of the Big Bang the universe dynamically emerged, its very nature controlled by a range of physical constants and the fundamental forces of nature.

These four fundamental interactions consist of gravitation, electromagnetism, the strong force of the atom and the weak force of the atom. We will better elucidate and relate them shortly. Pre Big Bang cosmology has been inconclusively treated (no matter how we try) so onto these constants that are believed to be universal in nature and holding set values in time.

Why and how are they the values that they are? They control the whole interaction of particles and strength of forces but they are the conclusions of observation with no available set of formulae to calculate them.

Further analysis is beyond our present scope and we mention a mere 5 of the given 27 recognized physical constants. We know c as the velocity of light, G as the universal gravitational constant and we have reinvoked Δ as the value of the energy density of the vacuum of space. We now necessarily introduce Planck's constant h as the quantum of action that is central to quantum mechanics and ϵ for Elementary charge or the charge carried by a single photon or electron. This describes the fraction of the mass of four protons that is released as energy when hydrogen is fused into a helium nucleus, the basic action of the sun and stars.

Addressing the Big Bang we continue a particular approach. Try a Jeopardy-like question: Exactly how miniature was the primeval atom to start with? Answer: Much smaller than an atom.

We revisit scientific tininess and the extreme brevities of time because we will need them in our wider theorizing. Recall how the ascribed Planck length represents a lower limit in the extremities of size. It is derived from three fundamental physical constants:

1) The speed of light in vacuum (c).
2) Planck's constant (h).
3) The universal constant of gravitation (G).

The Planck length is the postulated limit for the smallest distance having any meaning. Specifically it equals 1.616199(97) x10^{-35} meters and some 10^{-20} times more compact than a proton. It is the base unit in the system of Planck units and the time taken for a photon travelling at c to traverse the Planck length. This Planck time accordingly becomes the smallest unit of meaningful time.

As 10^{-43} of a second there are no smaller changes proposed and to illustrate this maximum brevity, direct measurements have not penetrated closer than 3.07 x 10^{26} Planck lengths. It is impossible to discern the difference between two locations less than a Planck length apart but this is not observable by instruments. Contemporary electron microscopes cannot get anywhere near. The Planck length does not appear on the regular metric scale and that nano second measurement of time of 10^{-9} of a second is 34 orders larger.

It is of no proven significance in physics but remains the smallest length possible in the interpretation of loop quantum gravity. Further, we conclude that at this scale the structure of spacetime is dominated by quantum effects and is probably quite foamy.

Volumetrically, more Planck lengths fit into a grain of sand than there are grains of sand in the observable universe. Note the tenet of Buddhist thought placing the smallest time possible as one in which a change could take place.

The cosmos came into being 13.798 +/- 0.037 Ga ago from a hot Big Bang derived from a singularity smaller than a proton. Let us agree that that the age of the Universe is the time elapsed since the Big Bang in the current observable universe. By IAU convention it is the duration of the expansion so far. I read that this was an equivalent 4.339×10^{17} seconds ago so someone must be counting very closely. How did so much came from so little and where it is all heading? Cosmology's first order of business is to introduce the standard Big Bang model.

The cosmological constant Λ now describes the vacuum, phantom or dark energy somehow inherent in the otherwise vacuum of space. It is no longer an abandoned bulwark against universally attractive gravity to maintain the static cosmos. This first intended then retracted by Einstein's self-confessed "greatest blunder" as expansion became the clear case.

This gives rise to the Lambda Cold Dark Matter concordance or ΛCDM model, the best-fit scenario of origin including a cosmological constant and cold dark matter. It safely assumes general relativity to be a correct description of gravity on the large scale and confidently incorporates both cosmic inflation and the vitality of dark energy, enigmatic as this last factor presents. It does not expound the conditions leading up to the Big Bang. We saw how these remain mysteries to both science and philosophy and that the former has not driven the latter into redundancy.

In this sense, philosophy (and its majors) are not dead but became strongly superseded by new interpretations, essentially rerouted through advanced empirical science. Cosmology became observational and the time is ripe, as Einstein once admonished, to rewrite philosophy in something more durable than honey. Our ancient Greek friends would love it.

Joseph Silk once observed how "humility in the face of the persistent great unknowns is the true philosophy that modern physics has to offer."

Let us not attempt any overly ambitious history of the universe but rather review some salient points from its past, present and future. The Big Bang was decisively not some small conglomerate of extant matter expanding in a destabilizing great explosion into preexisting space. Emphatically tiny as the initial state was during that proposed Planck epoch from zero to 10^{-43} of a second, the quantum effects of gravity must have dominated physical interactions. Hence, gravitation and all other forces must have been unified at the outset. There was no space for any differentiation.

In any realistic model we posit that space itself colossally inflated very early in the evolution of the cosmos. Inflation was an exponential expansion in the very early universe accompanied by quantum fluctuations on a microscopic scale that was magnified to cosmic proportions to seed its full growth and structure. We stick to this guiding principle.

It is therefore simplistic to regard the Big Bang as a massive explosion from a pre-existing primeval atom although the concept did serve an introductory purpose to describe some literal point of departure into a radically expanding state. Think on what we observe moving out in all directions then mentally run the action backwards like a movie. The hurtling galaxies must have been closer in the past to the point of a single unit. A much smaller entity from which everything sprung is implied.

Note that our model does not entirely *commence* with inflation. Not even in the extremely rapid action of the first phases of the Big Bang that we will summarize. After the Planck time, the hyperbrief period of grand unification or some equivalent describes the crossing of transition temperatures when the forces first separated. The inflationary epoch comprises the first part of the electroweak epoch swiftly following the grand unification epoch. Placing this era of uncertain duration for inflation as commencing 10^{-37} s into the prime origin and lasting until 10^{-33} or 10^{-32} s, this probably contributed most to the observed homogeneity of the universe and its apparent flat and isotropic state.

We propose three phases of development; very early, early and structure formation. Here is a synopsis as compressed as an event horizon.

Very Early Stage.

Planck Epoch or Era. From zero to 10^{-43} seconds.

We propose a gravitational singularity and that the four fundamental forces all have the same strength. It was possibly unified into one force held together by perfect symmetry. Temperature 10^{32} K.

Grand Unification Epoch. From 10^{-43} s to 10^{-36} s.

The force of gravity separate from the others that remain unified and the earliest elementary particles (and anti-particles) begin to be created.

Inflationary Period. From 10^{-36} to 10^{-32} s.

A rapid exponential expansion caused by separation of the strong nuclear force. The elementary particles left referred to as the quark soup is distributed very thinly across the universe.

Electroweak Period. From 10^{-36} to 10^{-12} s.

As the strong nuclear force separates from the other two, particle interactions create big quantities of exotic particles including W and Z bosons Higgs bosons. The universe still consists entirely of radiation but now supports mass.

Quark Epoch. From 10^{-12} to 10^{-6} s.

Quarks, electrons and develop and with the cooling the four fundamental forces take on their present forms. Quarks and anti-quarks annihilate each other, but a tiny surplus of the former survive, which combine to form matter.

Early Stage

Hadron Epoch. From 10^{-6} to 1 s.

Temperatures cool further allowing quarks to combine to hadrons like protons and neutrons and electrons collide with protons to form neutrinos. Temperature 10^{12} K.

Lepton Epoch. From 1 second to 3 minutes.

Charged leptons such as electrons and antileptons such positrons dominate the total mass. As they collide and annihilate each other energy is released as photons that in turn create more electron-positron pairs.

Nucleosynthesis. From 3 to 20 minutes.

Temperatures continue to fall and atomic nuclei begin to form as protons and neutrons combine to produce H, He and some Li. After 20 minutes, temperature and density drop to halt nuclear fusion. Temperature 10^6 K and falling.

Photon Epoch or Radiation Domination.

From 3 minutes to 240,000 years. Plasma as a hot and opaque soup of atomic nuclei and electrons and energy becomes dominated by photons.

Recombination or Decoupling. From 240,000 to 300,000 years.

H and He capture electrons in recombination. With electrons bound to atoms the universe eventually becomes transparent to light. These photons can now travel freely and are detectable as the background cosmic microwave radiation. Temperature down to 3,000 K.

Dark Age or Dark Era. From 300,000 years to 150 million years.

Between the first atoms and the first stars, activity had dropped dramatically and the universe is no more than mysterious dark matter.

Reionization. From 150 million to 1 billion years.

The earliest quasars form by gravitational collapse and their intense energy re-ionizes the surrounding universe. Here is the second major phase change for H since Recombination. Ever since the cosmos has consisted largely of ionized plasma.

Structure Formation

Star and galaxy formation. From 300 to 500 million years.

Gravity amplifies small irregularities in the density of primordial gas to form the first stars. Collapsing under their own weight this triggers nuclear fusion within short lived supermassive Population III and metal free stars of some M

100_{suns} (not yet observed). Their ashes from supernova events allow the further generations of Population II and Population I stars in discrete galaxies made up of groups, clusters and superclusters.

As indicated by theoretical particle physics, we calculate for an original state 100 million trillionth smaller than a proton with an accompanying temperature of 10^{34} K. All known laws of physics fail to apply. Particle accelerators can partly replicate the prevailing conditions but only take us so far in the comprehension of high-energy regimes. I personally refute that the Dark Ages refers to anything other than a lack of data.

All models are to some extent speculative to describe the initial scenario. They are only the best current deductions. Whilst specifics remain moot, there were certainly major and distinct epochs during the first tenths and hundredths of a second, corresponding to whole eras of change including the original separation of the principal forces. It is not possible to give a reliable radius in meaningful units to the newly hatched cosmos at any early stage.

This postulation of inflation now accompanies the standard model of the Big Bang, a concept principally due to Alan Guth in the early 1980s. It is the accepted best explanation for the origin of spacetime. In inflationary cosmology, we noted that the earliest meaningful time after the Big Bang is the time of the end of inflation generally set after the period 10^{-32} s. It was the most dynamic expansion the young cosmos ever underwent. By inflation, amplification by a factor of 10^{20}-10^{30} times possibly took place.

The obvious objection is that special relativity precludes travelling through space faster than c. By their own frames of reference distant objects are at rest in the space they occupy. "Through space" is the operative clause here because it does not constrain space *itself* from expanding far more rapidly. This is a critical concept. A beginning of time is tougher to grasp than particle-wave duality but we must go along with them until something better comes up.

From the very earliest Planck Epoch and after several intervening phases of astonishing brevity the size of the universe expanded exponentially. The duration of inflation is uncertain and it seems facile to state that it was long enough.

We are prone to saying that things are as they are because they were as they were for the fine-tuning of the universe.

In response to how there became something rather than nothing in the absolute beginning, we attempt the rationale that formative quantum fluxes instigated all of matter, time and space. Wait a moment, what are we proposing?

There cannot be anything that more severely strains credulity. The posit is that the sum total universe extending untold billions of light years originally derived from a super microscopic dot whence all matter and energy came. In any theater of superlatives, this takes the cosmic cookie.

Creatio ex nihio is Latin for "creation from nothing" and it merely sounds informed. It is as mind-boggling concept as a point source of practically infinite density, temperature and curvature to start with. Now add that it is expanding so vigorously that it might go on *ad infinitum* and we meet the same linguistics.

How can we grasp this? Some grand initial quantum flux is a real proposal to fix the huge conundrum where the rules of general relativity break down and we lose our scientific moorings. I am torn between cracking the champagne to celebrate progress in the area or drinking it anyway because we cannot know — an epistemological crisis. It is said that one does not study cosmology but rather it happens to you.

We do not conjecture that the cosmos always existed or clearly not in its present state. A hot Big Bang at a finite moment in time is overpoweringly evident from expansion, the CMB and the cosmic distribution of the elements. Some possible and alternate fates to the processes of expansion appear in the next chapter.

Another line of enquiry poses what is *observable* as distinct from what is *observed* in the visible universe. The observable universe is a spherical region incorporating all matter observable from Earth in the present due to electromagnetic radiation from these sources having time to reach us since the inception of the cosmological expansion. It is as far and as much as we *can* see. Not that we have identified and expounded everything in its compass by any means.

Any limits for an observable universe lie at a radius significantly greater than anything so far observed. Neither does the merely observable constitute any barrier or edge to the cosmos. We have not run into any limits so far and are unlikely to do so. The *visible* universe includes signals emitted since re-combination from the surface of last scattering whilst the *observable* includes signals since the commencement of the cosmological inflation.

According to the applications of general relativity, some regions can never interact with ours in the entire lifetime of the cosmos due to the finite speed of light and the ongoing expansion of space. Space may expand faster than light can traverse it and hence there are mega distant regions with which we can never interact. Different regions are causally disconnected from us or any other metagalactic vantage point. There must be numerous other visible macro domains depending on the observer's position, radically removed in distance from each other. Astronomers will never go out of business.

As for this expanding cosmic event horizon where the extremities of the observed universe progressively contact more and more of the observable as time goes on, the true and eschatological scale of the full cosmos cannot be assessed by present knowledge. The observable universe is no constraint on those unobservable regions beyond. Plugging such exponents into calcula-tors to get to quantitative grips can force them to "Error." Solid quantifica-tion is a good thing in all science so where does it lead before speculation loses all utility?

By inflationary theory the present universe could extend beyond the ob-servable by at least a factor of 3×10^{23}. This value is derived from when infla-tion began and we reason that its size was equal to the speed of light times its age. We can apply lower or higher estimates at will like a mathematician's playground.

Safely assuming that the current proper distance to the particles com-prising the CMB equals the radius of the observed universe, this gives an unconfirmed estimate of 45.7 billion light years in any direction. Recall that the visible universe includes signals emitted since the recombination event and the observable since the beginning of the cosmological expansion. The difference is since the Big bang in traditional cosmology and the end of the inflationary period in modern versions. Prospectively, the comoving distance to the edge of the observable universe is some 2% higher at 46.6 billion light years.

Despite an ever-expanding horizon the causal disconnections of the limited observable prevents signals from reaching us from outside it. This is simply because there has not been enough time for the waves to reach us. Nor will there ever be despite the observable event horizon continuing to grow outward in all directions.

Future technology may allow the observation of the proposed neutrino background that is older than the CMB or even more distant events by their gravitational waves. That we have now detected gravitational waves from lesser distances is again, very heartening. It promises a new era of what already is a Golden Age of cosmological discovery.

Perhaps Population III stars would be less difficult to finally observe but currently we can only see as far back as the era of photon decoupling in the recombination epoch and the surface of last scattering. Before then was full plasma opaque to photons.

Previous to introducing inflation it was tempting to place the whole shebang at a maximum 13.7 billion light years in radius as indicated by the age of the cosmos. Also to place a limit of the most far flung recessional velocities as very close to c and here lies the edge of the cosmos. Logically, here is all the time there is and there is as fast as anything can ever go. Is not the time elapsed since the starting point the veritable age of the universe and the distances traversed since that time its physical limit? Game over.

Emphatically, there are major flaws in this reasoning; bad enough to give totally misleading conclusions. The universe is not governed by flat space-time but rather is highly curved on cosmological scales. Because space *itself* continues to expand, the values for distances obtained by c multiplied by a cosmological time interval cannot bear any physical significance.

Neither on cosmological scales can we still advocate that the age of the universe equals the reciprocal of Hubble's constant H_0 (pronounced H naught) and that the reciprocal $1/H_0$ equals our long sought Hubble time or age of the universe. Some relationships stand between distances and speeds of recession because the further objects are receding at greater speeds but the calculations are complicated by general relativity, dark energy and inflation. It had been simpler with the "nearby" galaxies whose spectra Hubble successfully examined to first show the expansion. On very large scales there is significant departure from the linearity of the Hubble law:

$v = H_0 D$ where D is the proper distance of a galaxy.

Finally, it is an extremely remote possibility that the universe could be *smaller* than the observable. The most distant galaxies could be duplicate images of closer bodies, formed by light that has circumnavigated the cosmos. This is difficult to prove or even sample in any test because the different images of a suspected identical galaxy would be at different eras, rendering very different appearances in development. It is not a popular idea.

There are two other serious issues concerning the infant universe. The horizon or homogeneity problem poses how different regions of the universe that have never "contacted" each other due to massive distances inexplicably maintain the same temperatures and other uniformities of physical properties.

The visible universe is causally disconnected in a less profound way than the observed/observable dichotomies but the best solution is again, cosmic inflation and the later decoupling. The CMB is very nearly isotropic–the same in all directions. Its significance is its ubiquitous presence and we have dispensed with variable values for c or G over large time, as was once proposed. If the universe had started with even tiny different temperatures in different places the CMB would not be isotropic. Indeed, ripples and irregularities in it proved very influential to future development.

Place it as a kind of residual glow from all parts of the sky derived from the era of decoupling having been cooled by some 380,000 years of expansion to some critical level of density and temperature when the cosmos became transparent to radiation. We tell ourselves that there is no need of other mechanisms to even out temperature by the time of decoupling.

The second flatness or oldness problem arises from the observation that space on a cosmic scale seems almost exactly flat. The universe seems permanently poised on a knife-edge between expansion forever or a final recollapse in the far future and the current density is observed to be very close to a critical value.

The early cosmos must have had a density even closer, possibly deviating from the critical density by one part in 10^{62}. The curvature of the universe is dependent on the energy/matter density and once more it is the inflationary model coming to the rescue. It partly explains why the universe has an apparently flat geometry. As the universe expands, the curvature is redshifted away more slowly than matter or radiation.

Inflation stands on a triad of successful reasoning. Possibly solving the horizon, flatness and monopole problems. With the latter question, rapid expansion diluted their levels to unobservably low levels.

About the fine-tuning of the universe, even tiny deviations governing the fundamental interactions of particles would have produced major effects and whole differences on the emergent cosmos. Its character could easily have turned out very differently.

Major possibilities included an expansion of plasma that would never structure into stars. Very alternatively, a swift recollapse after the Big Bang due to more powerful gravity and attraction might have taken place. (See Martin Reis *Just Six Numbers*.) Extrapolating backwards we always encounter the fine-tuning problem. The contribution of curvature to the universe appears to be exponentially small, whole orders of magnitude smaller than the density of radiation with Big Bang nucleosynthesis.

By M theory other universes with alternate critical values among those constants not only could exist but do so in unseen plurality. They are forever hidden from our limited perception of three spatial dimensions and time and inflation also gave rise to them. These alternate dimensions are wholly too small to be assessed by direct measurement, permanently curled up in the folds of fundamental particles. The causation and role of dark energy are other vital studies in the future, as is the Higgs field. Possibly this not only gives mass to particles but controls the ordering of galaxies.

Turning to reference material, the highest confirmed distance by current knowledge (was) a galaxy where the spectroscopic redshift $z = 8.6$ corresponds to a "mere" 600 million years after the Big Bang. In revision, consider Abell 1835 IR 1916 with $z = 10.0$ and its light redshifted by a factor of 11. A very fortuitous gravitational lensing allows the image. This indicates an object seen 460 million years after the Big Bang when the cosmos was 3.5% of its present age. The possibility of this galaxy fragment 1/10,000 of the mass of the Milky Way undergoing vigorous star formation during the later Dark Age is the best interpretation. Other Hubble deep field images may have reached to $z = 12.0$ and the pictures are breathtaking.

It is fortifying to be using such terms of mere millions of years *after* the Big Bang rather than much greater spans of time *ago* from our vantage point. Further developments are strongly anticipated here, such as reaching far enough back to unequivocally view the earliest Population III stars, which we think were the first ones to form.

Thou shalt not get bogged in detail or scare readers with equations. It is time for a summary of our aims and to demand that any successful model must explain at least the following prevailing and established conditions:

1. The large-scale structure of galaxy clusters.

2. The existence of cosmic microwave radiation. This originated in a recombination era when the universe first became transparent to radiation and was subsequently red shifted down to a temperature of 2.73 K.

3. The distribution of hydrogen, helium, deuterium, lithium and other elements.

4. The accelerating expansion of the universe.

We are still reeling from the latter's effect on cosmology. The once sought "cosmic deceleration parameter" where some inevitable slowing down has long since vanished from the literature although it could have gone on in the cosmic antiquity. All related hypotheses should address both the origin and ultimate fate of the cosmos and there is no viable alternative to a formative hot Big Bang scenario in the primordial cosmic past and by consensus, an eternal expansion in the future. This evokes major discussions.

The first challenge is to ask whether it will all expand forever? We will attempt this formative question in the simplest terms and review other possibilities later. We could start by posing if gravity as attraction is ultimately that strong and far reaching as to overcome its apparent rivals driving repulsion? Was the Big Bang big enough to go forever? Perhaps the universe has not made up its impersonal mind about its own metafuture *i.e.*, the decision has not yet been made and the questions are wholly premature. From our perspective it is like asking for a reliable weather prediction for a day a thousand years hence. Here is the best we can do, assuming only two possibilities of ultimate collapse eternal expansion. We will consider refined details of these options as part of the section on the Density Parameter.

Take Ω (Greek capital Omega) as the ratio of the actual density of the universe to the critical and minimum density necessary for the universe to collapse under its own gravitation. Possibly the rate of expansion has not been uniform throughout full cosmic history, a consideration bringing us once again to assessing the part played by dark energy. On that, a riddle first appeared with the dynamics of spinning galaxies and galaxy clusters in 1933. Zwicky remarked on speeds faster than predicted for parts of them and speculated on something that did not absorb, emit or reflect light holding the Coma galaxies together. They did not rotate by Kepler's laws. In retrospect this "missing mass" problem as he put it was the first hint of the importance of dark matter. Meanwhile his colleagues demurred on any real significance in the dynamical studies that gave the first indications. A few years later Sinclair Smith found similar evidence in the Virgo cluster of galaxies and in 1940 Oort noted that 90% of the mass of the Local Group was dynamically unaccounted for. It was a long-standing problem.

Before we entertain further theory let us interpret some of the more extravagant formations that are observed. These are classes of objects as distinct from large scale anomalies like the Fermi Bubbles or Hoag's Object or mysterious individuals like Epsilon Aurigae, the Methuselah star or Tabby's star. (Do check progress on the latter case.) We must carefully incorporate bodies such as quasars, supernovae, neutron stars and black holes into our schemes. These are perhaps the strangest objects for energy output or stark individuality and are integral to what we see apart from the hosts of normal stars. Whilst billions of them compose the estimated two trillion known galaxies, any approach requires a brief assessment of the stranger entities that comprise the physical universe.

Exotic Beasts Of The Cosmos

Quasars

The first quasars were found in the early 1960s and due to their intense radio emission resembling point-like stars were dubbed quasi-stellar radio sources. We conclude that they are highly energetic discrete cores of galaxies forming long ago. They now include whole groups gravitationally lensed in appearance.

More common radio quiet ones had their names morphed into quasi-stellar objects or QSOs. A good standpoint is that quasars do not exist now; rather they *did* in a whole earlier and vanished epoch of development as the busy embryo nuclei of young galaxies with a supermassive black hole. Their high redshifts indicate general distances of 10 billion light years but the closest example 3C273 lies at 2.5 billion light years away in Virgo. (It is a target for good amateur telescopes.) QSOs might have been the very first structures to form and are thousands of times more powerful than the luminosities of regular galaxies.

They may have ended the proposed Dark Ages, operating between 150 and 800 million years after the Big Bang. They are intrinsically the most luminous objects observed yet are notably compact, some showing a variability over periods as short as days. This is a clear indication of low volume accompanying enormous energy density. In interpretation, quasars release trillions of times more energy than the sun emanating from sources on the scale of the solar system in volume. Over 200,000 QSOs are now known and the most compact are termed blazars. By consensus, it is purely our angle of view that distinguishes them from those or other distant radio galaxies where the jets of energy are not pointed in our direction. Accretion discs surrounding supermassive black holes are the accepted mechanism.

Supernovae

These are dramatically exploding stars consuming their fuel reserves in a busy few weeks at the last evolutionary stages of a massive star's life. They are prodigal suns more energetic than a nova that for such a brief time can outshine most of the others in its galaxy. We propose two distinct mechanisms for the causes of supernovae.

Firstly, the sudden re-ignition of nuclear fusion occurs in a degenerate star. Here, the gathering of material from a binary companion through accretion or merger hugely raises the core temperatures. For a second cause, the abrupt gravitational collapse of a star's core can trigger a similar action. Supernovae are classified with subdivisions according to their light curves and absorption lines in spectra including Type I lacking hydrogen lines and Type II showing the Balmer series. Diverse supernova remnants or SNRs are widely on show like the Cygnus Loop. Such débris result from throwing a wild party before collapsing into long term retirement as spinning neutron stars or humble white dwarves.

Three supernovae have been seen in the Galaxy in the last millennium. These are the Crab supernova of AD 1054 and Tycho and Kepler's "new stars" occurring four centuries ago. Other stellar upheavals produce what are termed cataclysmic variables and there is a class of 10 known recurrent novae in the Galaxy. Stating that two occur per century in the Milky Way is clearly a lower limit to real occurrence but we observe about 150 a year in other galaxies. Having forged heavy elements in their great cauldrons of nucleosynthesis they serve important functions by enriching the interstellar medium. Their shockwaves similarly contribute to the future production of generations of other stars. A supernova's death throes can eject material up to 10% *c*, including the building blocks of life and organic compounds.

Neutron Stars And Pulsars

Spinning neutron stars are the remaining cores of previously massive stars and generally the final stages of very advanced but not complete collapse. They are generally the central remnants of supernova explosions, now comprised by implosion of highly compressed neutron degenerate matter. From "pulsating star" we adapt "pulsar."

Their estimated density is 10^{14} g/cm^3 with micro and millisecond pulses directionally released through intense magnetic fields, somewhat like a lighthouse beam in rapid spin. Pulses last on the order of microseconds with a typical interval of 0.25–2 s between them and the directions are fortuitous from our standpoint. Masses of hot gas blast away from the original star, propelling the outer parts of an expanding SNR as the material diffuses into space. For a mass 1.5 M_{suns} the radius of a neutron star is about 30 kms. They are generally the remains of a supernova Type II but possibly form when a white dwarf undergoes significant transfer of matter from a close companion star.

Neutron stars represent the densest states still visibly glowing because their collapsed cores are still radiating. They are the hardest objects, more compressed than diamond with an active core as small as a few miles. Zwicky and Baade had proposed such formations of a density alike to neutrons in 1934. Pulsars are fast spinning and highly magnetized neutron stars.

Some older ones may have been spun up and not slowed with the loss of rotational energy, likely due to transfer of matter from a companion. They may occasionally originate from white dwarf bodies rather than a nova phase having formed less explosively. The closest example is 280 light years away in Cetus with the catchy name 1RXSJ141256.0+792204. The first one discovered (Bell and Hewish 1967) was half seriously dubbed LGM 1 for little green men. We wish...

The Crab Nebula

The first optically identified pulsar and the first associated with a known prior supernova remnant was located in the Crab nebula, the huge and active remnant of the greatest supernova in recorded history. Coincidentally, it was the opening entry on Messier's catalogue as M1. As a searingly bright source seen in Taurus, it may have been naked-eye visible for as long as two years. Native American art depicts the event and there are written records from Chinese, Japanese and Arab sources. Inexplicably, there is no mention in European annals. John Bevis first observed the Crab nebula telescopically in 1731. At a distance of 6,500 light years its present dimensions are 13 light years across and still expanding at 1,000 miles a second with filaments in the temperature range 11,000–18,000 K. It is now some 70,000 times the solar luminosity and remains the strongest permanent radio source in the entire sky. The Crab pulsar contains 2 M_{suns} compressed into a diameter of 28–30 kms.

The source emanates pulses of radiation in 3.3 milliseconds slowing at 28 nanoseconds a day. Its periodicity is so precise as to rival atomic clocks and allow the calibration of X ray detectors. The Crab pulsar's relativistic winds generate synchrotron radiation producing the bulk of the overall emissions. The most dynamic feature are zones where the equatorial winds collide with the surrounding nebula. There is a saying that there are two types of astronomy: the astronomy of the Crab nebula and the astronomy of anything else. Surprisingly, the higher the mass of a neutron star, the *smaller* the radius — because gravity pulls it in ever more tightly. Further process of implosion could cross a certain threshold where the strength of its gravitational field no longer allows radiation to escape. No, not ever. It is bounded by an event horizon to form singularities known as:

Black Holes

A black hole can ensue at the core of the terminal collapse of a very massive star. They may be typical at the centers of galaxies, which are certainly the first parts to form and hence the oldest parts of galactic structures. Like my readers, they have collapsed as individual bodies and are no longer *directly* observable by any means. The gravity well is so great that escape velocity exceeds *c* and no radiation or information can escape back out to normal space. The event horizon marks a point of no return, permanently encompassing a region forever turned away from the normal continuum into its own Schwarzschild radius. They are the final phase of collapsars, reaching maximum entropy and density where as a singularity the mass distorts normal spacetime around it. It is the last word in independence for the very densest of all cosmic matter and speculation rages on conditions within them and other peculiar properties like their spin, charge and temperature. Are such things knowable? Here lie dragons.

The Schwarzschild radius for the earth is about one inch so imagine a state so dense that the total mass of the earth is compressed into a sphere that compact. Matter genuinely exists in these conditions. These advanced objects have undergone such implosion as to form an event horizon and pass the point of no return to the normal spacetime continuum. As to observation, they are sometimes detectable by *indirect* means. Their presence is inferred by the behavior of companion stars and detaching matter around the extreme gravitational field. Mini black holes must exist. Finally, black holes ain't so black. The quantum conditions of hyper slow evaporation and quantum effects giving rise to Hawking radiation suggests a hypothetical emission beyond a shrinking event horizon but an evaporation to zero is at a pace comparable with the long-term outcome of the cosmos.

We can spend our fleeting days of rational consciousness entertaining some powerful ideas on the nature of the universe and bequeath much to future generations to better expound. There is still much to explore; realistically, we have only scratched the surface and a proper grasp of cosmology, relativity and quantum theory are at the cutting edges of human understanding. Shall we pause for some interpretation and an update on Haldane's standpoint? Krauss puts it well:

The tapestry that science weaves in describing the evolution of the universe is far richer and far more fascinating than any revelatory images or imaginative stories that humans have concocted. Nature comes up with surprises that far exceed those that the human imagination can generate.

— *A Universe From Nothing*. Preface p xvi.

That author has emphatically little time for an omnipotent deity or any effective explanatory power associated with traditional religion. Modern cosmology brings a unique sense of proportion to an informed 21st century worldview, one that leaves the old paradigms behind.

Let us embrace the idea that along the way the greatest thing we can experience is the mysterious. When Einstein commented that the good Lord is subtle but not malicious, the former remained strongly in favor of the scientific determinism that Newton had perpetrated. He did not approve of the direction quantum theory was going at the time, specifically the belief that the speed and position of a particle cannot be both fully known. In criticism of the Uncertainty Principle, Einstein objected, "God does not play dice." It proves that He is an inveterate gambler who throws the dice on every occasion and still has many a trick up his sleeve. Einstein was in favor of strict determinism.

For one, gravitation retains enigmatic. Having said that, we are no longer entirely stumped by classical physics to explain action at a distance. By the presence of mass, gravitation is a distortion of spacetime. This is one difference between the classical and relativistic forms of physics, where gravity was semi-successfully treated as only a force. We are within the gravitational field of the sun and considering the solar wind also hits the magnetospheric and atmospheric influence. Hold to John Wheeler's statement: Mass tells spacetime how to curve and spacetime tells mass how to move.

Black "holes" are not an absence, as the name might imply, but rather matter existing under the ultimate state of compression. Their densities make white dwarf stars look like snowballs.

Quark stars are hypothesized as the middle ground between the neutron star state and becoming a full-fledged black hole where a proposed quark broth reigns as their composite material. They are likely extremely rare (one hesitates to declare them nonexistent) with one candidate 450 light years away.

We know of some 2,000 neutron stars in the Galaxy and the time lapse images of the maelstrom that is the Crab pulsar are staggering to view. A star requires at least 8 M_{suns} to finally become a neutron star or prospective black hole.

The original term "dark star" makes a better description with Wheeler coining "black hole" in 1949. The term "frozen star" is more apt but "dark star" is very descriptive. It originated in 1783 with John Michell writing for the Royal Society on imagined stars of such density that no light or information can arrive from them. He mentioned that they might be indirectly observable by the effects on double stars of which they were members. This was farsighted thinking and Laplace entertained similar ideas a few years later. For names, anything is better than the risqué French translation "trou noir," and I gather that the Russian (pronounced "chernaya dyra") has a similar problem.

Gravity reigns so dominatingly that escape velocity exceeds c and cannot be attained. He-that-gave-us Relativity originally interpreted them as purely hypothetical and even disproved them from actual existence on one mathematical outing. Such singularities definitively exist.

As for the strangest anomaly, frozen time rules within a black hole as the weirdest feature of all concerning their departure from normal spacetime. Time does not pass within a black hole. I still cannot wrap my head around stationary time as a *de facto* concept.

In the process of formation neutron degeneracy pressure can be critically overcome by the intensity of inward pulling gravity endlessly collapsing the body. This is not the case with normal stars entering the phase with or without a supernova precursor. Such dense states of matter are not how electrons in shells about normal atoms and the forces maintaining them generally exist.

Jumping into a black hole would make for a very long weekend. There would be no way back out again. Not that any creature could survive the g forces of entry, lovingly known as the spaghetti effect of being strongly elongated by radical acceleration by steepening gravity. From achieved spaceflight we conclude that speed as such has no bad effect on astronauts but acceleration, deceleration and prolonged weightlessness do.

One basis of stellar evolutionary theory is the condition that massive stars propel through their life cycles faster and we have suggested that some singularities were primordial in the phase of Structure Formation.

Did primordial black holes soon form from the earliest Population III stars? What part might they play in the long-term future of the cosmos? There will be a suggestion here.

Gravity is the weakest but by far the most operative of forces over the larger scales of distance if not within the smaller ranges of atomic structure. We mentioned that we had once anticipated a cosmos inexorably slowing down due to that universal gravitation of unfailing *attraction*. Our thinking never fully departed from Newtonianism in making this a feature that must eventually occur but there was never any evidence of a hypothetical cosmic deceleration parameter.

On the contrary, the opposite effect of acceleration was eventually found. Bear with me again to appreciate a piece of guesswork.

In historical context, about five years after *Principia* was published, Newton received a letter from a minister named Richard Bentley, pointing out that attraction and hence a collapse of the stars was inevitable under his universal gravitation. In his reply, Sir Isaac was minded to an infinite uniform collection of stellar bodies so they would not "fall down in the outsides and convene in the middle." If every star were pulled evenly in every direction then stability would result.

He was never comfortable concerning gravitation acting at a distance and also wrote back that gravity must have an agent acting constantly according to certain laws whether material or immaterial. He could briefly wax as enthusiastically as any scholastic on the infinite power of God.

The Rev Bentley still has a point. One prior cosmological view is that universal collapse has to happen in the long run due to the long, long arm of gravity. The present could lead to an eventual turn around for the furthest-flung galaxies and eventually a major reverse in direction. Expansion could therefore be merely a primary phase for a lengthy but limited period including the present and foreseeable future. It was even suggested that the value of G was dropping off but like any variable speed of light over time these ideas were entirely dispensed with.

Here it is then. Assume that gravitation finally wins. I regard black holes as being the most up market bodies of all in relation to the full cosmic future. Meaning they have already arrived at where everything is heading like the vanguard of an imploding universe reversing into a second great phase of *contraction.* They have already completed their part of the great cycle of time and space.

This is part of the case for the collapsing, oscillating, bouncing or cyclic grand state but in the light of an accelerating universe such predictions are forcefully deferred.

Away with theory a moment, it is time to address some straightforward questions about singularities and event horizons. This inevitably crops up because black holes do seem to fire the popular imagination, as does interstellar spaceflight and alien life. I am snobbish about SF plots but it is good that these are in popular consciousness.

Where in space lies the nearest definitive example of a real black hole? Similarly, what is the closest pulsar or rotating neutron star to the solar system in corroborated observation?

Bring on the first contestant in tonight's "spot the black hole show" and we'll think up a suitable prize.

The binary star cataloged as A0620-00 or the same variable star designated V616 Monocerotis consists of a K type main sequence star with an unseen partner of calculated mass 6.6 M_{suns} that is too massive to be a neutron star. Two separate X ray outbursts in 1917 and a better-observed episode in 1975 suggest a pulling of matter from the K type star onto an accretion disc. With a period of 7.75 days an ellipsoidal shape for the visible component is safely assumed to be affecting the visibility of the surface area and changes in apparent brightness.

During the latter outburst tracked by the Ariel 5 satellite A0620-00 was temporarily the brightest X ray point source and regarded as an X ray nova. At an unconfirmed distance of 3460+/- 390 light years this represents the closest known black hole to the solar system.

We also put forward the case evidenced by the X ray observations of the suitably named Cygnus X-1, a possible black hole, lying 6,100 +/- 400 light years away. By its galactic latitude and longitude this is inward to the Orion Spur near its approach to the Sagittarius Arm of the Milky Way.

It was first noted in 1972 following the strong X ray source that was discovered eight years earlier by a rocket-borne collection of data. Observed as a guest star over a period of forty-four days in AD 1408, some Chinese astronomers were hardly aware that they were witnessing the formation of a black hole.

The blue supergiant O type star HDE 226868 possesses about 19 M_{suns} and a surface temperature of 31,000 K. It is locked in a close orbit about an invisible companion. The latter's mass of 6 M_{suns} exceeds the limit at which a collapsed object must become a black hole and the largest possible neutron star cannot exceed 3 M_{suns}. Cygnus X-1 itself does not send out bursts. The displays of X rays come from the superhot material taken regularly from the supergiant onto an accretion disk centered on the dark primary body.

We are confident about a hot inner region with high ionization and a plasma state out to a less intense and lower ionizing outer region extending perhaps 5,000 times the Schwarzschild radius or about 9,000 miles. It is in the inner region that matter is heated to millions of degrees over a relatively small area and by its known type the visible star is not capable of producing such a copious X rays flux.

Calculations for Cygnus X-1 indicate a black hole of approximately 14.8 M_{suns} with a Schwarzschild radius of about 28 miles. Extensive observations down to milliseconds and observed intervals reveal variability and intervals indicating a thin and flat disk of accreted matter. As seen from Earth, the system does not eclipse and the inclination of the orbital plane has been narrowed down to 48° +/- 6.8°. At a separation distance of 0.2 AU and an orbital eccentricity of 0.0018 its companion follows a nearly circular orbit over 5.6 days.

The system retains a common motion through space of some 13 mls/s within a stellar association called Cygnus OB3. If a supernova was previously associated then the remnant would more likely have been ejected away from the system. The conditions of them remaining suggest a star of originally 40 M_{suns} rapidly evolving then collapsing directly into a black hole without such an explosion.

A little do it yourself calculating is useful for understanding Cygnus X-1. Do not fear mathematics because it is a dynamic tool.

For the figures of a body 28 miles in radius and calculated spin of 800 times a second:

Circumference = 2πr

2πr = 176 miles for the circumference of Cygnus X-1

c = 186,000 miles per sec

(176) x (800) = 140,800 miles per sec for its rotational speed

140,800/186,000 = 76% c

The spin speed for Cygnus X-1 is therefore 76% that of light.

A third case of GRO J165-40 exists as a black hole candidate. It comprises an F type primary star and an unseen companion with an orbit of 2.6 days but at a remove of 11,000 light years it lies almost halfway to the galactic center. Its high velocity of 112 km/s suggests a supernova origin for propulsion.

This trio of black holes indicates their known distribution in at least this part of the Milky Way but it is severely limited. By their interactions we would have located any others lying nearby by now. Apart from singularities including small ones at far greater remove many others are extant in the Milky Way. There must be many more lone examples not observably interacting with a binary companion that remain unknown.

The successful detection by LIGO of authentic gravitational waves represents a new level of observational sophistication. Reflect on releases of energy from merging black holes or neutron stars so great they send ripples through spacetime.

New Dangers. Earth In Peril!

We will not be foreseeably assailed by a black hole but what of the newly unveiled potential disasters cosmic science has revealed? As if earthbound fires, earthquakes, tsunamis, floods, disease and war are not enough, pollution is now identified as a major culprit.

As for self-inflicted agonies, if we really were as stupid as to wipe ourselves out, it would be a mass species suicide unheard of in nature. In some unseen future, the epitaph thus would be titled *Homo asininitus* in an alien's report on vanished life forms. It would read: "Sol III, an emerging technology neglecting the basics of self-preservation. Such unplanned over proliferation is unusual for a species employing single reproduction and lengthy gestation periods."

We now know of inherent dangers from space as well. Safety is a relative thing and there can be no guarantees of survival, however smart we get. We will phase into a discussion of the newly recognized hazards via some extreme renditions of more familiar disasters. Earth could prove a harsh mother indeed by serving up turbo driven versions of old tricks.

Firstly, it is clear that climate change will present many challenges. With the rise of sea levels and temperatures, we may well be in for at least changing patterns of precipitation and the expansion of deserts in the subtropics with greater warming expected over land than oceans.

Whole streams and patterns of current are changing, with a maximum over the Arctic as glaciers and the regions of permafrost continue to retreat. Episodes of extreme weather including droughts, hurricanes and intense rain are already happening more often. Greenhouse gases like carbon dioxide, methane and nitrous oxide remain in the atmosphere for a long time and the overall inertia of the global climate system indicate shifts for possibly thousands of years to come. From the Intergovernmental Panel on Climate Change (IPCC) Fifth Assessment, 2013, we read that the global surface temperature is likely to rise a further 0.3° to $1.7\ ^\circ C$ in the lowest emissions scenario and 2.6° to $4.8\ ^\circ C$ in the highest, and human activities "have been recognized by the national science academies of the major industrialized nations and are not disputed by any scientific body of national or international standing."

In addition to over population and the challenge of feeding us all, serious climate change might be enough to overwhelm humanity's ingenuity. Unless something substantial is undertaken the future looks daunting like nothing man has faced before.

Apart from a massive dose of common sense acting on our common plight I can only weigh in that a series of technological fixes might mitigate matters. Can we think our way out before it is too late?

Could we find a way to redress the balance of the atmospheric gases or find a means to reduce pollution *and* cope with humanity's sheer numbers? We must put our best minds and means to tackling the problems before they engulf us all. We must realize that there may not be much time.

Geophysically, take an extreme version of earthquakes. The asthenosphere could reposition itself and shift parts of the greater crust and sea floors above it, driving crustal movements into some sort of overdrive. This could set loose a tumult of seismic upheavals, supervolcanoes and shake-ups that have been building for Ka before suddenly being released. Mega tsunamis could ravage the planet and active fault lines miles in width could smash civilization in hours; our cities laid waste with devastations setting back centuries of progress.

Imagine the whole west coast of the United States pitching into the Pacific Ocean or the state of Florida entirely engulfed by the surrounding waters that wait no longer to encroach. The seas could radically reposition themselves as the whole oceans reclaim the face of the world with millions killed in hours.

The continents could disappear beneath the tumultuous waves leaving only the highest regions and mountains above the new eustatic level. Mother Earth might shake us off by a combination of natural volcanic fire and engulfing water, her whole face again liquid "from sea to shining sea." It is reassuring that such geophysical extremes are unlikely, as far as we know.

Meteorologically, the atmosphere could stage a paroxysm of upscaled tornadoes and hurricanes the size of small nations crisscrossing the world. A hydrologic cycle gone awry could cause rain for months on end in some places or little precipitation in others, disrupting whole patterns of agriculture. Following this, the world's temperatures either zoom to hundreds of degrees of heat or plummet to dire cold for keeps. As organisms we are not naturally designed for anything but a small range of temperatures and a major shift would soon destroy us.

What of our protective layer of gases? A vital component is the ozone layer lying 12 to 19 miles above the earth in the lower parts of the stratosphere as a naturally occurring shield from harmful ultraviolet light emitted by the sun. We have only known about it since 1913. O_3 is a highly reactive oxygen molecule and if it were to be wiped out or substantially depleted the results would be catastrophic as unabsorbed and dangerous radiation increasingly reaches the earth's surface. The atmosphere has changed in the past and could do so again. Now consider some newfound menaces from space.

Nearby Nova

A nova or supernova would have to be relatively close, in the ballpark of a few tens of light years to adversely affect us with dangerous waves of energy. These could be more than radiant heat although this would certainly be enough to badly rock the planetary boat. I've waded through studies concluding that a Type II supernova would have to be in closer proximity than 8 parsecs to destroy 50% of the earth's ozone layer by incoming radiation and fortunately, there are no close candidate stars.

The variable red supergiant Betelgeuse may take this explosive course over the next ten million years but this star is some 700 light years away. This is far enough not to be a menace but such is its size it permanently delivers a shock wave 2 light years long. Has a civilization ever been destroyed by a nova from its parent sun? Somewhere in time and space this must have taken place.

A less philosophical question is this: Is our grasp of astrophysics really that reliable to say that no star within a few tens of light years is predictably going to blow itself up? Being overdue for a naked-eye nova might not be so bad.

Rampaging Black Holes

A small black hole could distort the orbits of the planets, altering their established paths as it ran amok through the solar system. The planets would be like tiny corks at the mercy of a giant gravitational wave if one literally invaded our space, propelling the inner worlds into the depths or hurling them into the sun. Any refurbished orbit or shift in Earth's orbital eccentricity could give us the worst of all worlds of high and low global temperatures. Radical freezing and alternate heating could ensue so swiftly that not even the insects could adapt in time. Any such change would be apocalyptic.

Worse, an interloping rogue singularity could shift the planets away from the sun entirely, transforming the earth into a free nomad planet wandering in space. Alternatively, a black hole in rapid passage could cause a more slow motion set of disturbances among the paths of the planets. A readjustment of their orbits and reshuffling gravitational resonances could ensue over indefinitely.

How about a passage so disruptive as to break the sun itself up into a latter day star cluster? Or it could swallowing it whole along with its retinue for lunch? A mini black hole would make a threatening asteroid look positively benign and there would be scant chance of (indirectly) seeing it coming.

An Upheaval Involving The Sun

We have already touched on the life of the sun itself and what would happen if it significantly cranked up or rolled back its level of energy output. It is obvious that we owe our continued existence to stabilities descending from the solar constant but it is not popular knowledge that the sun is already a slightly variable star. Reconsider the surface environments of Venus or any outer planet for the ranges of temperatures in the planetary neighborhood.

If the sun changed the earth could be suddenly thrust into great heat or somber cold by either cataclysmic heat waves or devastating new ice age. The latter could equally be caused if the solar system were to pass through thickening interstellar dust or diffuse nebula, cutting down the amount of solar heat reaching us. Next time you observe the solar apex, think what might lie in its direction.

We also saw that in any case of major upheaval by the sun there would be little we could do to counteract terminal catastrophe for all life on Earth. By fire or ice, the game would be up if we assailed on such overwhelming scales.

Gamma Ray Bursts

A big GRB within a few hundred light years is among these lurking doomsday fates our forebears never knew about. It would be devastatingly hot if brief in passing with a major onslaught vaporizing the atmosphere and oceans with powerful waves of radiation. In short shrift the world would become one lifeless hot desert. Again, we can relax because GRB sources are cosmically removed but are such threats really so implausible? We return to hard facts.

A GRB, equivalent to putting planet Earth in a global sized microwave oven could have been the active cause of a previous mass extinction in Earth history, prior to the asteroid wrecking the ecosphere of the planet of the dinosaurs.

The Ordovician Event some 400 Ma past may have wiped out 70% of marine species. It was the first of the five big five Phanerozoic events and such an energy burst places significant threats to planetary biospheres up to an estimated distance of 6,500 light years. Let us think about GRB more closely. A mere ten seconds from a close source would strip the atmosphere of ozone.

Gamma rays are the highest frequency, shortest wavelengths and most energetic of all electromagnetic radiation. Note that the potential such hazard WR104 in Sagittarius is thankfully 7,500 light years away. This Pinwheel nebula has a component Wolf-Rayet star predicted to become a core collapse supernova with the possibility of a long duration GRB.

At visible magnitude 5.8 the great GRB 0830319B of March 2008 known as the Clarke Burst was briefly naked eye visible in Bootes though I cannot trace that anyone actually caught sight of what was the furthest the human eye could ever have seen. It comprises the intrinsically brightest event ever observed, courtesy of the Swift Orbital Observatory and ground based cameras, which imaged the afterglow. We barely understand how a release of energy, focused in a beam and guided by high magnetic fields at outset can be equivalent to the sun's entire lifetime output.

From a source 7.5 billion light years away, the GRB swept through the solar system among many other regions of the past, present and future. It was, is and will be visible across half the observable universe. By the speed of light and the scale of intergalactic space it occurred (for us) when the universe was less than half its present age. Call it a discrete occurrence plucked from the full-scale cosmic landscape.

Asteroid Strikes

For statistical likelihood, asteroid and meteor strikes are far more a concern than hyper energy releases or gross gravitational interferences. Some impact is inevitable to a greater or lesser degree and the subject is well covered elsewhere. A common sense approach soon shows that serious impacts have been mercifully rare in recorded history. We enjoy the celestial fireworks of meteor showers as long as long as they keep to the size of pebbles burning up tens of miles aloft.

The legendary temple at Delphi might have contained a sacred meteorite and other solemnities of veneration are in the record, probably including the

Black Stone sacred to Islam. Not until the early 19ᵗʰ century was it finally es- tablished by the French Academy of Sciences that stones can indeed fall from the sky and that meteorites are extraterrestrial in origin. Thomas Jefferson was wrong about this but reviewing celestial mechanics, nothing falls on the earth. They fall toward the sun and the earth gets in the way.

The Woodleigh crater in Western Australia may be linked to the Perm- ian-Triassic mass extinction of 250 Ma past and of course, the Chicxulub crater bears witness to the impact that spoilt the dinosaurs' day. Under ideal circumstances small meteors from predictable meteor streams can be observed impacting the moon but it would be fun to see a big one hit *there* if only for old time's sake. The regular Orionid meteors are fragments of Hal- ley's comet and many regular streams have been identified to such parent bodies or origin.

Here is a curious blast from the recorded past. Take the report of a pecu- liar celestial sighting a passage in the *Historical Works Of Gervase Of Canterbury* for the year 1178 AD. In their own words some 12ᵗʰ century monks saw the crescent moon "split in two" with accompanying "fire, hot coals and sparks."

Did they witness a real impact event and even the formation of the crater Bruno? On closer analysis, it was not a lunar feature being blasted into exis- tence but instead they had the singular sight of a large exploding meteor in their direct line of sight to the moon. The lingering smoke of an atmospheric fireball explains the blackish appearance of the moon they describe.

That particular crater is geologically young but far older than nine centu- ries. Had this been something leaving an identifiable feature, it is likely that débris thrown out would have given our monastic brethren and all others a grand meteor display in the following weeks. It is not in the annals that any such fragments showered the earth so we conclude that even on this rare occasion, no such things came to pass on Earth as it was in heaven.

How near or predictable are NEOs? We'll consider another example in the future tense.

Asteroid 99942 Apophis is expected to pass within 20,000 miles of us on 2029 April 13 and should be 3ʳᵈ magnitude visible. We have already calculat- ed for the keyhole path it thankfully avoided on a previous close call on its 324-day orbital period and the impact risk is very low. (See "Torino Scale" for danger assessment.) It still spun off papers with titles like the universe's most dangerous asteroid. Its path is unlikely to shift on its subsequent pass in 2036.

Apophis is some 1500 feet in diameter with an average orbital speed similar to that of the earth's and in broad figures its orbit has several factors in common with the earth. Some other small but speedy NEO sneaking up on us is unavoidable. Spend a few reflective hours with http://neo.jpl.nasa.gov/orbits to assess what we really know and study intersecting orbits in the comfort of your own friendly blast proof shelter.

An apocalyptic asteroid strike wiping most life from the land and seas is possible at any time, a particularly cheery thought. We're confident by now to have the means to see a big one coming. The speed, angle of approach, mass and density of the interloper comprise the many variables for the degree of possible damage. Hitting the ocean would create the mother of tsunamis. Going out to get it preemptively with suitably armed spacecraft has spawned several mediocre movies but a few decent "day after" SF novels are worth the read. (The Niven–Pournelle *Lucifer's Hammer* is one of the best of this sub-genre.)

See Duncan Steel's *Rogue Asteroids and Doomsday Comets* to rationalize your general fear factor here or books on the Tungus event. In 1908 we dodged a major bullet and we may not be so lucky next time. Coming merely hours later the Tunguska cosmic body might not have dived into the Siberian wilderness but have hit St Petersburg or Paris.

Principles And Forces

In the human past we were oblivious to the menaces from space we have described but Mankind has always been in the grip of great natural forces. Placating them is some component of primitive religion but suffice to say it is only of late that have we accumulated any decent understanding of them in scientific terms.

The corollary is that planet Earth is not the center of the universe and moreover, there is no such focus for the wider cosmos. On cosmological scales this merely appears so due to the continued expansion. From its standpoint any galaxy could construe itself as the hub with everything else speeding away.

This is the corroborated view for which we first thank Nicolas of Cusa in the 15th century and Giordano Bruno for the power of rational thought in interpreting the starry sphere as a host of other suns and planets. We must also pay homage to Copernicus, Tycho, Kepler and Galileo for widening our view in processes and legacies that were not without serious travails. In a way, the battle lines with religious authority were drawn up in the Renaissance as the Mother Church faced challenges on new fronts both sacred and secular.

Luther had more impact on society than he originally aimed for. As a pious man, he first sought reform from within for the Church, not to fundamentally challenge it or set up some rival Christian faith. He was methodical and reasonable in his writings, despite his reportedly neurotic character, but his masterstroke was having the printing press on his side.

Henry VIII sought a divorce rather than breaking with Catholicism and setting himself as the head of a new Church. But, no matter how he tried, he was doomed to fail in his duty to leave a brace of hale, hearty and unquestioned male successors.

The Reformation saw much intellectual energy directed into theology, but this inevitably struck at the very power base of the Church. The results included shameful episodes like the trial of Galileo and the execution of Bruno, the martyr of heliocentrism. The Church was in no mood to tolerate further dissent, including a sun-centered Copernican cosmos. The human force of ignorance can be so dark as to obscure the enlightened.

In our day, Arthur C. Clarke once wrote that if there are any gods whose chief concern is man, then they are not important gods themselves. He also reported that he had no religious beliefs but the concept of a God fascinated him. It is strange how the universe laid on a big GRB at his passing. The two events cannot be related after 7 Ga of transit from its source, but what a send-off he received in that phenomenal coincidence of timing.

The Earth is not an object but a series of interrelated and highly active ongoing processes and we only exist with their cooperation. A particular advance on our appreciation was the iconic "Earthrise" photo from Apollo 8, which became the picture of the century. In going all the way to the moon it was more impressive to behold the sight of the earth in space.

The Copernican Principle

The Copernican principle (Bondi) holds that the earth holds no privileged position. This is equivalent to the broader Principle of Mediocrity that is "heuristic in the same vein."

This is probably most meaningfully approached in considering that galactic address for the solar system in better detail. Any misjudgment that the sun is in some privileged position within the local Milky Way has long been was laid to rest by observation. It had been an assumption from Herschel up to the early 20th century. Let us ask *where* we are in our Galaxy, our individual island universe among a known 2×10^{12}. I guess we do need a "trillion" as an arithmetic unit.

Our knowledge for the precise layout of the Milky Way is yet again, rather poor but slowly improving. Here is the classic case of not being able to see the general shape of the galactic wood for the vast numbers of starry trees that block the full perspective. (Run through Scorpio or Sagittarius with binoculars or a low power and you get an inkling of the center of the Galaxy.

The best descriptions place us 27,000 light years from the center in what is likely a barred spiral galaxy of type SBc with two major arms comprised of great filaments of stars. Results from the Spitzer space telescope suggest this shape for the Milky Way and spiral barred galaxies are a common structure among so many others.

The galactic arms surround a central bulge that forms a flattened spheroid about 8,000 light years across its poles and about 8,000 light years for the diameter of its equator.

Of the three types of galaxy that are spiral, elliptical and irregular there are further classifications giving rise to the tuning fork diagram. Our Milky Way is of intermediate size and nucleus with loosely bound spiral arms.

This SB class is plentifully seen with the sub classifications a, b and c rated to describe the descending order of tightness of the associated arms. The original classification of galaxies had been a bit rushed.

In our case the Scutum-Centaurus and Perseus Arms have the greatest observed densities of both young bright stars and older red giants. Two other minor arms, Sagittarius and Norma are filled with pockets of young stars.

The sun lies near a small partial arm called the Orion Spur or Orion Arm probably located between the Sagittarius and Perseus Arms. Its individual motion is 20 kms/s towards the solar apex in Hercules but the much greater Milky Way rotates over a galactic or cosmic year requiring 225–250 million years. It has made about 16 full journeys around since the solar system formed or a third of a rotation since the end of the dinosaurs.

The Cosmological Principle

The Cosmological principle (Milne) notes that on a sufficiently large scale the properties of the universe are broadly the same for all observers wherever placed. Generally the cosmos is homogeneous and isotropic — looking largely the same from any vantage point. The "horizon problem" will be declared solved by inflation.

The Cosmological principle therefore has its roots in *Principia*, Newton speculating that the universe was infinite and homogeneous so as to prevent some great gravitational collapse under gravity. The spatial distribution of matter in the universe appears homogeneous and isotropic in the big picture. There are only limited observable irregularities in the grand structure. So far, however, we interpret the course of its physical evolution since the Big Bang.

Mach's Principle or Mach's Conjecture

Local physical laws are determined by the large-scale structure of the cosmos.

Local inertial frames are determined by the distribution of matter on a cosmic scale.

The philosopher Berkeley expressed the basic idea and there are no less than eleven variations descending from Bondi, simplified to "mass out there influences inertia here".

Gravitation

All the forces were presumably unified at the outset. Gravitation is the weakest force by far but has the furthest reach. Newton showed that a point mass attracts every other point mass with a force directly proportional to their masses and inversely proportional to the square of the distance between them. The universal constant of gravitation G did not emerge in physics until 1798, a full 111 years after Principia and seven decades after Newton's death but now quantify gravity as having only $6 \times 10^{-39 \, lt}$ of the strong force. It proves to be a distortion of spacetime by the presence of mass and in full expression, $G = 6.674 \times 10^{-11} \, m^3 kg^{-1} s^{-2}$

We accept that as an independent force gravitation first separated at an immediate 10^{-43} s into the Big Bang, the first meaningful moment.

The Strong Force

This is responsible for maintaining the nuclei of atoms because it keeps quarks in hadron particles such as protons and neutrons together, which in turn bind to form atomic nuclei. It is very short ranged, operating over distances over the diameter of a medium sized nucleus, typically 10^{-13} of a centimeter. It is generally attractive but can be effectively repulsive at the levels over which it operates. At the range of 10^{-15} meter (1 femtometer) it is 137 times as strong as electromagnetism, 10^6 as strong as the weak interaction and 10^{38} the force of gravitation. It probably differentiated during 10^{-35} s.

The Electromagnetic Force

In causing electric and magnetic effects and can be either attractive or repulsive, acting between electrically charged particles. It is much lower in strength than the strong force but hypothetically of infinite range. If the strong force is assigned strength of 1.0 then the electromagnetic force is a comparative 0.007. Diversification of the electromagnetic and weak forces occurred around 10^{-35} s.

The Weak Force

This controls radioactive decay and the behavior of neutrinos and has the shortest range of all, entirely within the atom. It operates over a mere 0.1 % the diameter of a proton and is 10^{-6} times the strength of the strong force.

The Higgs Field

It is moot that this should be ascribed to the list of fundamental forces. The Higgs field is an energy field thought to exist everywhere and it is accompanied by a fundamental particle called the Higgs boson, which the field uses to interact with other particles in the process of electroweak symmetry breaking. The Higgs field is not a gauge field because it does not have a gauge symmetry associated. If a fundamental force is defined as any action mediated by any kind of boson then it qualifies as a force of nature. When the temperatures were sufficiently high that electroweak symmetry was unbroken all the elementary particles were massless. At a critical temperature the Higgs field became tachyonic and the symmetry was broken by condensation as the Higgs bosons acquired mass.

The obvious question is how gravity can be justifiably called weak? It is ever present as the all-pervading universal force, ranging wholly beyond falling apples and the moon in orbit. Famously, seeking their connection was something of a point of departure for Newton and all he gave us.

The answer is that gravitation of the earth is very low and the reasoning goes thus: In consideration of an escape velocity one can temporarily defy it with a jump in the air and achieve very temporary weightlessness whilst the free fall of attraction is not opposed. This lasts but a moment but a mere 7 miles a second serves to permanently break free of the greater field such as a probe launched out of orbit. This is a remarkably modest velocity compared to the earth's speed around the sun of 18 mls/s and others motions it possesses.

Additionally, the general movement of the sun and its retinue solar system toward the solar apex and the spin of the greater galaxy carries our world faster than the speed equal to its own escape velocity. The point is that the earth's escape velocity is low compared to other values of speed inherently associated with it.

This provides a handy demonstration of the scale of the gravitational field produced by a body of the earth's total mass. Quantitatively, the value of acceleration for a falling object in its immediate environs is measured as 1 ms^2, defined as 1g. Both the earth and moon are comparatively dense planetary bodies and we see how much mass spread over what volumes engenders such modest gravitational fields. For the moon it is 1/6 g.

The sun's attraction is greater by a factor of 28 and Jupiter's is 2½ that of the earth. This allows us to draw clear and distinct ideas of the masses required to engender their gravitational fields. In the big picture it is a question of the distances gravitation operates over and compared to the other fundamental forces its great reach makes it the dominant one.

Let us further muse for what other outcomes there could have been — had those values been only slightly different for the fundamental forces. Another value for the constant for the strong nuclear force within the atom could have eliminated the production of carbon and other elements in the hearts of stars.

A difference in the weak nuclear force could have affected the radioactive decay of neutrons and the cosmic abundances of the elements would have strongly been compromised from what is observed. Considering the electromagnetic force, the sizes and fundamental structures of atoms would have caused a different set of rules for all molecular chemistry. Most importantly G being otherwise would have dictated much for the life cycles and behavior of stars including nucleosynthesis.

We do not know why the constants and forces of nature behave as they do but we safely assume they were determined in the Big Bang. Any Grand Unified Theory necessarily takes on an explanation for this and here is one of the barriers to constructing the Theory of Everything. Things could easily have been significantly different, especially with G. It is no longer debated that these constants of nature slowly change or adopt different strengths at remote cosmic distances. (Refer to the discarded 'tired light' hypothesis.)

No matter how strong or weak you take your Anthropic Principle it is a colossal fluke that we are here to investigate and appreciate it so let us seize the moment to focus our scientific energies on the greatest enigmas of all.

There was no hint of any of this until our era. Whilst these are among the findings of advanced physics, taking us beyond the perfection näively ascribed to nature they put an entirely new spin on the Argument by Design i.e., the world has design so there must be a master designer. None of the arguments were conceived with any knowledge of nuclear fusion, black holes, accelerating universes and Goldilocks zones in mind. In a way it is a violent universe and way removed from the sedate gentility the mediaevalists asserted. It does not "only go to show."

There were apparently huge choices in precisely how to build the universe. It could have been quite different in whole tenor and possibly they simultaneously *are* if one opts for M theories. It gives a whole new meaning to the plurality of worlds; one on a higher level to its previous meaning as asserted by a few luminaries of our intellectual past.

Yet in all cases, what there was prior to the inception of the Big Bang remains a matter of profoundly speculative philosophy. It seems an impregnable barrier to knowledge.

The new evidence of acceleration indefinitely postpones any such discussion of the cyclic, rebounding or oscillating universe. We now favor that the initial great expansion did overall slow until perhaps 6 Ga ago. This was until dark energy took over and speeding up again prevailed, representing over 50% of full cosmic history since the transition set in. We do not agree what the current rate of expansion is and the rates may continue to vary. If some kind of energy is somehow inherent to space itself then as space expands dark energy will increase and maintain the same density instead of diluting out as its overall strength falls. If only we knew.

Let us restate current opinion. Some slowing down may well have gone on during the great past like the first 8 Ga but this was superseded by newly accelerating expansion. Most encouragingly, the forthcoming Dark Energy Spectroscopic Instrument Survey (DESI) is under construction with the Mayall telescope at Kitt Peak. It aims at a 3D cosmic map to study in detail the expansion history and physics of dark energy.

In the drive for original thought there has not yet been a suggestion that the universe might seriously fragment with some parts slowing to reverse and others set loose forever. The universe could reach stages of semi collapse and semi indefinite expansion.

Would this be consistent with the proposed dark energy takeover only affecting the outermost observed regions? I find the proposition curious by absence that the universe could enormously fragment with an outer region expanding endlessly with another separate section recollapsing.

The WMAP probe and newly gathered Plank microwave map of the entire sky reveal irregularities in the CMB of the very early universe. Looking back, these slight concentrations of matter led to the formation of the first galaxies like seeds growing to full growth over monumental time. We propose that following the early universe and the steps of structure formation, basically galaxies, blossomed by a bottom up formation.

We'll shortly visit the possible fates for the universe in the ultra future.

Any objections to a Big Bang never fused into a coherent alternative option despite the efforts of this now retired Continuous Creation or Steady State hypothesis. It seemed a rational alternative with matter needing only to be newly created at an astonishingly low rate to keep things going. Observational evidence entirely points to a colossal explosion and ongoing expansion.

Neither is there any grounds to propose any series or plurality of lesser Big Bangs also taking place and we conclude there was a single grand event in the formative past rather than explosions anywhere else. Perhaps we should not be so hasty to disclaim dense dark energy expanding into vast empty space before shrinking and pinching off separate baby universes.

Those pictures revealed in the Hubble Space Telescope deep field imagery are astonishing for the depth, clarity and sheer quantities of external galaxies. In studying the earlier state of the universe and building a picture for the development of the cosmos we literally are receiving information from fundamentally younger phases of developments and events from the cosmically remote past. Recall the best case in point: quasars *existed* in the past as a vanished epoch for the development of galaxies. We are only just getting the news.

Such are the matters landing squarely on our scientific doorstep. The best available data for a recipe for the universe currently indicates:

< 1% Visual matter (stars, nebulae etc.)

22% Non-baryonic dark matter (exotic unknown particles)

4% Baryonic dark matter (protons, neutrons, atoms, molecules

< 5% Baryonic matter (including intergalactic dust, brown dwarves)

~ 73 % Dark energy

The Density Parameter

Ω is now more fully described to apply to the average density of matter in the universe divided by a critical value for that density. This latter may be sufficient *or* insufficient to eventually halt expansion, not that "weighing" the cosmos is any feasible feat. The fraction of the effective mass of the universe attributed to dark energy as described by the cosmological constant universe is denoted Ω_Λ

Note that the definitions are tied to the critical density of the present era. The critical density presumably changes over time but the energy density due to the cosmological constant remains unchanged.

The overall geometry of the cosmos refers not to the local curving of spacetime but the context of a closed, open or flat universe. We can posit a different final outcome from the shape of the universe and assume that its shape plays that crucial a role in how things turn out. A positive vacuum energy resulting from a cosmological constant implies a negative pressure and *vice versa*.

If the energy density is positive, the associated negative pressure will drive an accelerated expansion of empty space. Another important hypothetical ratio is an equation of state *i.e.*, the ratio of the pressure that dark energy puts on the universe as the energy per unit volume. It would determine how the dark energy responds to the great expansion.

There are three broad scenarios for the shape of the universe and the relation of Ω to unity, a quantity divided by itself yielding of course 1. Weighing the universe makes the Labors of Hercules look like child's play. However, we currently place a conjectural value of 0.3 for Ω, tending to expansion. We shall first consider the options presented by an open universe before we tackle a flat or closed arrangement.

1. Open Universe

> If $\Omega < 1$, then the geometry of space is open and negatively curved like the surface of a saddle. Gravity does not halt the expansion and by the critical factor of dark energy the expansion continues forever.

2. Flat Universe

> If $\Omega = 1$, then the geometry of space is flat. The average density of the universe equals that critical density allowing a flat universe

to endlessly expand but at a continually decelerating rate. This assumes that dark energy is not dominant. Should this mysterious vacuum energy prove decisive in the sense of taking over, the situation becomes the same as the open universe.

3. Closed Universe.

If $\Omega > 1$, then the geometry of space is closed like the surface of a sphere. Gravity eventually halts the expansion and the beginning of overall contraction ensues. A new singularity will ultimately be produced in the process of gnaB giB which is of course, the phrase Big Bang in reverse.

Density is therefore crucial to any postulations of final outcome or to borrow a theological term, the eschatology of the universe. It may be overly ambitious to sketch meaningful values for the density of creation and the long-term effects of gravity versus expansion enhanced by dark energy. However, the trend of opinion is more consistent with indefinite expansion placing us in a one off metagalaxy and a single grand performance. Is the universe open or closed? Can it achieve its own escape velocity? Here goes. Should the whole shebang reverse into contraction the closed universe rationally heads towards a:

Big Crunch: 10^{11} years from now—

Broadly, this is a symmetric view of the Big Bang's final outcome. Matter and spacetime recollapses into a reformed dimensionless singularity with unknown quantum effects requiring major consideration. The average density of the universe proves sufficient to halt the expansion, pause for some unspecified period then commence contraction. The process might not be so easy. It could include active stars so close together that they cannot export their internal heat as the temperature of the CMR built up and the stars morph into a hot and strongly heterogeneous gas whose particles would break down before being absorbed in coalescing black holes. Along the way we have a gnaB giB with clever astronomers discovering that everything is contracting at an increasing rate. This could lead up to a speculative:

Big Bounce

Following the Big Crunch, the universe again assumes a micro particle of infinite mass and density. This is also termed the bouncing, cyclic or oscilla-tory universe, where the cycle of repetition from the collapse of the previous incarnation repeats with a new Big Bang promoting a new expansion.

The collapse of one instigates the other, with the possibility existing that the most recent Big Bang derived from the former implosion of another pro-cess in reverse, time after time. We could be living in an infinite series of performances and we wonder how long to the next turnaround. The pause between the two events of recollapse and new expansion could 10^{-43} s or Ga. There is no frame of reference but do note that black holes have already have arrived at the stage of total collapse.

Now we turn away from trampoline universes as the concept of repeated expansion and recollapse. Should the cosmos continue to expand indefinite-ly, we must consider several different alternatives for an open grand state. There are plenty of possibilities. Extreme long term projections for this openness include the:

The Big Freeze: 10^{14} years from now and the related Heat Death10^{150} years from now—

> The cosmos grows dark: Existing stars finally die with no new ones created. The universe reaches a high entropy state consisting of a great bath of particles and low energy radiation, eventually reaching full thermodynamic equilibrium. The cosmos tends to lower temperatures and approaches 0° K. The universe advances to a state of maximum entropy with everything evenly distribut-ed with no temperature gradients. (Big Freeze in the case of un-ending expansion could occur in a flat or open universe.) With a positive cosmological constant pulling everything inward again a Big Freeze could be the outcome even in a closed universe. Note that this cessation of energy is consistent with any shape for our models and envisions a scenario of final temperature minimum. Maximum entropy has been reached.

Big Rip: 2×10^{10} years + from now—

> Here, imagine a runaway expansion where the rate of accelera-tion by the supreme thrust of dark energy entirely wins out. The cosmos has spread too far for gravitation to control by attraction

any longer. All material objects are dismantled into unattached elementary particles or radiation and the dark energy density and rate of expansion become infinite. Even black holes will have the time to evaporate (as is foreseeable with other models' projected time scales) and the universe reaches a final end state of particle singularity. The electromagnetic forces holding atoms together and even atomic nuclei are torn apart in this mega remote epoch an the cosmos finally winds down. A domination of unbound elementary particles finally results, the fundamental structure of matter rent asunder by the overwhelming vacuum energy.

Multiverse Or Meta Universe

It is popular speculation that our beloved Big Bang created other effects than what we unquestioningly interpret as a single universe. If so, the Big Bang was merely one among many unobservable effects and unseen dimensions are truly extant. Sometime in its rapid and energetic birth or shortly following the constants were assigned other values and the interrelation of forces was formatively different. This suggests a multitude of other levels as the ontological case. (In philosophy *ontologia* is the branch of metaphysics dealing with being.) A chain reaction myriad set of multiverses, matter and anti-matter, quantum foam, Higg Bosen particles and the whole panoply of M theories are there to ruminate upon. Parallel or real alternative planes of existence could abide, about which there is no cognition.

False Vacuum

In quantum field theory a false vacuum is vacuum existing at a local minimum of energy and is not fully stable. This contrasts with a true vacuum, which exists at the global minimum and its configuration is very long-lived and metastable. When a vacuum reaches its very lowest energy state it could tunnel into a lower state of energy called a vacuum metastability event and the transition to the true vacuum is stimulated by the creation of high-energy particles or through quantum-mechanical tunneling. A false vacuum approaching true vacuum in this way could split into new scenarios of forces and all resulting structures as the universe loses its previous characteristics. Speculation is becoming idle. The possibilities appear endless and our project of evaluating the long term fate of the cosmos turns into an intellectual game.

Can we simplify matters? Is the universe open or closed and how we might address the specifics? Speculations are predicated on the overall density of the universe with Ω related to the curvature of space. Recall that it represents the average density of the cosmos divided by the critical energy density dictating an arrangement of closed, open or flat. The consensus is for an open and eternal expansion and that all events since the Big Bang form a one off performance. Will we ever really know?

A finite beginning but an infinite future presents. There is a certain *fin de siècle* with cosmology pushed to this limit and perhaps we try to project matters absurdly far. It is hard to see how calculations invoking 10^{14} Ga hence can be valid. Let us remain optimistic about new data, although I fear that it will be a long time before we can understand it all in meaningful terms. Perhaps some friendly aliens could bring us up to speed one day with higher-level knowledge about the cosmos! Until then, I'm still voting for an oscillating universe because I back gravitation for the outright winner in the absolute fullness of Time.

Even tiny variances in those physical constants derived in the initial conditions of the Big Bang might have instigated different courses for the emerging cosmos. Or failed to maintain it for long. If universal gravitation were that part stronger a temporary expansion leading to a swift recollapse before galaxies formed could have been the case. The Dark Ages might never have progressed into Structure Formation before gravity took it all back over a far shorter lifetime than the present age of the universe. Conversely, an overall weaker universal gravitation presents a significant alternative. With insufficient attraction to pull anything discrete together like stars, matter would have expanded as oceans of elementary particles from the Big Bang and stars never formed.

Opting for multiverses, some versions of M theory includes virtual particles coming in and out of very existence in bubbles of quantum foam. In the litany of fundamental physical constants there is certainly no agreed value for D, the number of spatial dimensions. It is difficult to see how any breakthrough can be anticipated in direct support of M Theory and there are better things to concentrate on. This could my greatest blunder of judgment but I see better avenues of enquiry and more tangible pursuits in cosmic science. Squeezing out a desperate laugh the universe is open Mondays through Fridays but closed on Saturdays and Sundays.

CHAPTER 4. LIFE AND ORIGINS

In The Beginning

There is a saying that the most incomprehensible thing about the universe is that it is comprehensible. Primarily, it teaches us that we are part of an interconnected web of existence and we have done well in a short time. Whilst we cannot be the only creatures imbued with rationality we are undoubtedly low in its galactic hierarchy.

The conclusion is that we possess brains composed of matter promoting some element of consciousness. Nature's gifts to us include these tools of enquiry, the sum total of human perception, knowledge and imagination. For exploring inner space, the best definition of mins is our own metaphysical construct of this awareness. Plutarch (1st century BCE) wrote that the mind is a fire to be lighted, not a vessel to be filled yet it is sobering how little is understood of the functioning of the human brain. It is probably the apotheosis of this planet but we must guard against arrogance.

For any purpose attaching to cosmic nature there can only be the most profound wonder. It certainly justifies exploration. Asking why there is a universe might not bear any meaning and like Newton with gravity our best empirical descriptions fall far short of explanation. Perhaps shaking the tree of cosmology hard enough will allow some apples of wisdom to drop. Meanwhile, ponder what levels awareness can aspire to when fully allowed to bloom.

Let us focus on life's origin, or biogenesis, more closely. NASA's definition of life is straightforward — a self-sustaining chemical system capable of Darwinian evolution. This is succinct to the point of cryptic, but all attempts at devising a formal statement set off lengthy descriptions. Any definition must expound the differences between the organic and inorganic, and this is not proving easy.

Better is something along the lines of "organisms composed of cells that can maintain homeostasis, undergo metabolism, adapt to their environment and respond to stimuli, and crucially, grow and reproduce, and over great time, develop substantially further in the diversity of whole species."

Water, energy and carbon-based organic chemistry are essential conditions on Earth and we tend to assume that they are indispensable factors for life anywhere. Frankly, this is all we can do. A fluid medium of stable temperature, the triple valence of carbon and the inventiveness of proteins characterize the given parameters, yet such recipes may indicate a serious lack of information on our part. Perhaps we are mistaken to think carbon so essential that we tend to "carbon chauvinism." Silicon is a conceivable alternative, but carbon is the most adaptable for forming versatile compounds.

In exobiology, there are limits in extrapolating from the single example; it's like having only one piece of a presumed giant puzzle. This sets the bounds of rational speculation, so we turn again to our literary exponents to describe alien lifeforms, where extraterrestrials warp drive their starships from one episode to the next. Larry Niven's aliens are truly so and push the boundaries well. They are among the best you can meet in this esoteric art form, fictionally finding the "heartbeat in utero" of astrobiology. Less well-produced stories make their Earthling audiences laugh in the wrong places.

Aiming more at education than entertainment, these are the basic questions:

> 1. Is it easy for young biospheres in other star systems to nurture the progressive building blocks of life?

> 2. Can it only happen in the form of proteins commencing as chains of amino acids?

3. Can we assume that conditions are often ripe for intelligence and sentience to further raise itself to the heights of self-conducted evolution?

We have to start somewhere, so let's assume for the sake of argument that the masterstroke of the universe so far is giving rise to biological consciousness. For those not convinced, a counter case is presented.

Planets as potential ecospheres are proving to be common, and nature seems versatile enough to conduct any number of experiments in the drive of upward progress. There is always time. Observing how far its evolutionary paths travelled from unicellular organisms to a brontosaurus, much led up to the dinosaurs before the whole production was shattered.

We could be taking the first steps for a self-aware universe and from different lines of arrival, any other beings that might exist must encounter the same position. After all, we share the same physical universe and hold scientific laws in common.

So we consult two Nobel prizewinners on the matter. Having achieved so much on the structural and functional organization of cells, Christian de Duve considered it inevitable that life should arise, given all the right ingredients and the passage of time. Jacques Monod, who worked in deciphering genetic codes, reached a different set of conclusions, placing complex organisms as a gross rarity and a bizarre fluke.

So we draw a certain line in the scientific sand and introduce a crafted compromise called the Rare Earth hypothesis. This idea maintains that the generation of life and biological complexities requires an improbable combination of astrophysical and geological conditions. Accordingly, it hardly ever occurs and sophisticated life is an even scarcer phenomenon. Suppose Fermi is more right than wrong about Earth's uniqueness and there are only a few hundred advanced technological civilizations in the entire Galaxy. (I refuse to go lower than that!)

A Rare Earth equation has even been constructed as a riposte to the Drake equation, pointedly not factoring in complex life reaching these higher levels because such is too improbable to be considered. Rare Earth also demands tectonics, the influence of a large moon, a magnetosphere, lithosphere and oceans and other "evolutionary pumps" like a big guardian planet. In our case it is Jupiter and all this is deemed essential to the production of eukaryote cells as the starting point.

The facts of the case include that simple life soon emerged on the earth. Therefore, levels as low as bacteria and single celled organisms must be comparatively quick to rise in biospheres. About the soundness of this logic, we need to pause again. It is questionable that there is such a thing as "simple life" and until we discover how self-replicating molecules come into existence, the term seems an oxymoron.

Having done so by processes unknown, does life always drive to greater complexities, including the huge specialization of cells into organs? Might not the most modest versions remain forever unchanged as lowly reproducing microorganisms? We seem to have the best of all possible worlds for enduring abundant bacteria *and* the advanced diversities of animals and plants.

We have mostly agreed that there are three domains at the root of the "Tree of Life." These are Bacteria, Eucarya and Archea, all hailing from a common root. There is the remote possibility that the original two were preceded themselves and that the Tree was actually a forest, but biochemically we opt for a common and single origin. Very different species share remarkably common genes. Believe it or not, butterflies and whales are not that different in the sense of their genetic codes.

There are other appropriate analogies about biogenesis. Fred Hoyle once said that obtaining even a single functioning protein through the chance combination of amino acids was as likely as a star system full of blind men solving Rubik's cube simultaneously. He also compared the odds of a functional creature popping up to a whirlwind passing through a junkyard and constructing a jetliner by chance.

In a way, unraveling the huge evolutionary past is opposite to the conjectures of exobiology, the great unborn science. It must be easier to understand what has already taken place than speculate on the possibility of life beyond Earth, leaving us with evolutionists overwhelmed with information compared to frustrated astrobiologists starved of any data. There is much to probe right here, without sending vehicles off to Jupiter and Saturn, laudable as all such missions are. In the next decades we hope to properly address the possibilities that some forms of *elementary* marine life abides on the outer satellites of the solar system. Europa, Enceladus and Titan are the prime candidates in the biomes of subterranean oceans, such as they are. What major surprises await us in these exotic locations? The slated Europa Clipper mission to Jupiter's moon is designed for the purpose.

Possibly, life arose here further back than 4 Ga, and in the purely temporal context, what else was going on at the time? We could ask whether it began on Earth and possibly Mars before, during or after the Late Heavy Bombardment. Precisely what sort of enablement rôle did that episode play during the storm of material in the inner solar system 3.8–4.1 Ga ago? Like Shiva, the LHB could have played the parts of both creator and destroyer.

For example, did the LHB promote biogenesis by introducing prebiotic material or, alternatively, could it have violently snuffed it out only for the processes to recommence in a process of stop and go? On which side of the tailing-off of the impacts did it originate on Earth? Did they have any effect for stirring it up or was biogenesis a matter for the future when things had simmered down?

The precursors of life could well have been imported; or they could have sprung from inside the very crust of the earth. This latter scenario is asserted in the Deep Hot Biosphere theory. Never underestimate the power of groundwater. Swirling primordial soups and idyllic watery pools are equally viable sites for origin, and we can only rule out spontaneous generation, which like alchemy was ignorant of the basic laws of chemistry.

It is always hard to prove a negative, i.e., that something outlandish did not take place— although we generally write off panspermia or any deliberate seeding by outside intelligences. In our age, supernatural divine sparks are an unacceptable whitewash to science. By whatever means, the earth was only ½ Ga old at the time life first arose and it is a direct question whether it preceded or followed the formation of the moon.

We cannot attempt any abbreviated biography here, and the fossil record speaks cryptically and often in riddles. What's a few billion years between friends of common genes, as we jump ahead in the story to one notably productive period? We have ascertained that the Cambrian Explosion about 541 Ma ago commenced diversification by an order of magnitude, expanding to produce most of the animal phyla existing today. It compares to an orchestra striking up successfully with no rehearsal and no game plan devised in advance. With the great lizards, it could be said that evolution eventually got seriously carried away and was only halted by random interference from outside.

Every big book on astronomy or biology contains words to the effect that "life must surely exist elsewhere," and the following section is one opinion, which I believe to be rational. No special insight is claimed.

Life seems more likely to have arisen out of a specialized series of occurrences than a singular event. I find it hard to assume an entirely individual common ancestor on the molecular level at merely one site. It seems improbable that biology could reach the levels it has from merely one locale, as there are too many barriers and disruptions. This said, we saw how the entire physical cosmos energetically emerged from a supremely tiny source in the Big Bang, roundly presenting us with the ultimate in tall stories as accepted fact. Recall that remark of Darwin's about how daunting it would be to find the origin of either life or matter.

Accordingly, I propose that life arose in numerous places at about the same time on Earth and posit multiple geneses in a special set of conditions. Furthermore, everything here originally came from the solar nebula, so we must evaluate "indigenous" as meaning formed along with the primeval Earth, as distinct to "imported" at some later point during the early history of the solar system. There are plenty of organics in comets and the great molecular clouds in space. I suggest that some of these prebiotic materials were incorporated into the primeval Earth. (And presumably elsewhere, although no further action got going on the other planets. Or, possibly it did but failed to survive due to further changes in environment.)

There is of course no standard model for the origin of organic molecules, but there is a clever word. For the study of how life arose from inorganic material, we use the term "biopoiesis" which, like "adaptive radiation" in evolutionary biology, makes it sound we know what we're talking about.

So here is one suggestion, if you prefer not to bathe on radioactive beaches or crawl out of a primordial soup. We think the oceans formed 4.41 Ga ago, and fossilized microorganisms found in hydrothermal vents indicate an age for them — possibly as long as of 4.2 Ga in age. Life definitely began in the marine environment, possibly in those remarkable crucibles acting as the natural test tubes of biopoiesis.

As for that "life as we know it" phrase, the parameters prove substantially greater than previously thought. We find that life exists in extreme terrestrial environments including these black smokers and other unlikely environments. Conditions of both high and low temperatures, pressures, low oxygen, zero sunlight and pH levels long thought unaccommodating to life have been found to have thriving organisms lately called extremophiles.

So far we have stuck to cellular biological life emerging in a great ocean and/or little pools. At length it adapts and colonizes the surface, "adaptation" being the motherlode of unknown mechanisms. Now evaluate the staggering diversity of life on our own planet to project the limitless possibilities on distant worlds. While you are at it, conjecture further on the possibility that intelligence could exist out there in forms other than the purely organic. Machine and artificial intelligence, cybernetic organisms or other non-biological sentience such as energy beings could flourish somewhere. In all the depths of space and time, the bias of selection has no set boundaries.

As they say, curb your enthusiasm. It is merely a point of view that life and intelligence are matters of importance. Consider one extreme counter case. Perhaps biological intelligence is not so special.

I do not advocate the following but we seek a balanced view and we must consider different angles. According to the stringent methods of analytical science, a good theory should be falsifiable in sowing some seeds of its own downfall, i.e., specify where its fatal shortcomings might lie. Equally, furnishing successful predictions strengthens a testable theory.

Obeying this we generally prefer theses like "On The Supreme Importance Of Intelligent Life," but here is one alternative wholly disparaging the concept. Let us counter the assumption and assert that life is highly over rated. From the last words of Socrates in Plato's *Phaedo*, one learns that his at least is a disease from which he is shortly going to be cured.

Propose the following for life-bearing planets: Among the unused gas and dust left over from star formation are comparatively small quantities of matter permanently reduced to cool fluid states. The material is quite unable to produce energy by fusion and deteriorated into cold undistinguished satellites around their embryo suns. Further decomposition left them so inert as to form solid surfaces with thin gaseous envelopes and occasional oceans of inert methane or water either at the surface or shortly beneath. Their differentiated crustal tops, deeper mantles and warm cores characterize differentiated failure compared to the processes within the most modest brown dwarf star.

Perhaps life is an insignificant biochemical aberration among wholly degraded states of matter divorced forever from the high energy reactions that is the occupation of stars. Our locale forms a typical example, where relatively little material exists in this inactive state compared to the bulk of the atomic pile comprising the sun. It represents the literal star of the show and protein chains are nothing compared to proton-proton chains of fusion.

On, then, with our antithesis rationalizing the unimportance of life. Living cells are merely desperate viruses striving for full metabolic independence, enacting the poorest substitute for nuclear fusion where the real action is. Cell metabolism is a poor substitute for the dynamic and versatile processes going on in stars because they are negligible in energy production.

Biological activity is hence no more than a runaway pathogen amongst conditions so inert that nuclear reactions cannot take place, solid matter being completely degenerate. It is the lowest state of matter and only plasma as the highest is relevant to the destiny of the universe. Sterility is the natural order for the first 10^{10} years following the Big Bang until plasma intelligence arises. All prior illusions of consciousness are futile — if for no other reason than that organisms are far too short lived.

Recommended is Clarke's short story *Out of the Sun*, which inspired this ramble.

We now return to established science. About Jupiter's reputation as a failed star, it would have required greater mass by a factor of 75 to ignite by a Hydrogen Flash and the smallest brown dwarf stars are probably 30% greater in radius. The planets all hold some internal heat sources with this factor most pronounced in Jupiter. For reasons not entirely clear they radiate more energy than is incident upon them with the apparent exception of Uranus. All things considered it is not good deduction to underrate life so glibly but Asimov did point out that the solar system consists of one central star, two major planets and some débris.

In the earlier reducing atmosphere it was possibly electrical activity like lightning that catalyzed the creation of the basic small molecules of life such as amino acids. The slow oxygenation of Earth's primeval atmosphere was a major preparatory step to allow full bio diversification including photosynthesis and cell metabolism. Whether information or metabolism came first with proper cells is a dilemma. Researching how critical this point may be shows that it depends who you ask.

There is the tenacity and resourcefulness of life and that remarkable factor of zero waste and 100% recycle in its sum total activity. Nature's engineering is far better that anything we have ever designed and we cannot be justified in the denigration of intelligence as a result. A sterile universe is analogous to a magnificent opera in full flight with no audience to bestow any sense of appreciation.

Without an appraising intelligence it all appears spectacularly pointless until observers arise like a phoenix from the stellar ashes, immediately seeking funding to make sense of what has gone before. If only we had a hint on the commonality of even the humblest life anywhere else we could assuage some better sense of proportion.

For now we must content ourselves with the conclusion that we are made of star stuff and that everything making us up was originally formed inside stars long ago. This alone is a worthwhile conclusion. We will shortly review these processes of nucleosynthesis and the production of elements. This is how anything other than H and He came to exist beyond the Big Bang and what those silvery points in the night sky are really about.

Throughout the cosmic past these stellar factories were turning out the elements that eventually composed our whole physical and chemical environment. We concluded earlier that we are one outcome; a strange concentration of the heavier elements from generations of star birth and death in which their products and by-products eventually did an incredible thing. Obviously, we are one result and as the seats of our own consciousness this is how we got here after long preparatory processes.

As we saw, the earth was formed among a group of other small planets among the inner peripheries of material in the contraction and construction of a normal star from a huge primordial cloud of gas and dust. This much seems common, planetary systems being regular sideshows of stellar evolution accompanying their formative stages.

The evidence suggests that it is a regular type of gestation by the contraction of matter accompanying proto suns. We estimate that 80% of young stars may possess such temporary circumstellar discs of material to do the job and we will review the processes shortly. Note how we ascribe planets as a general early development. There must be different cases where much older stars give rise to planetary attendants later in life, the hypothetical Phoenix planets developing in the disks of dead stars.

There are numerous sites of star and presumably planetary formation on observational view but things do not happen so fast as to draw firm conclusions. Planetary accretion could proceed much slower under lower temperatures, less dense matter and strengths of gravitational fields among both old and young stars. Planets might not get going until later than the star's initial formation we only assume that the local planets are entirely of uniform age.

For those impatient for conclusions, I offer this perspective so far: We live beneath a thin, gaseous and protective canopy of a biosphere surrounding a small, rocky, curiously watery small planet revolving around an average star in an insignificant backwater of a Galaxy. It produced us and nurtures us. The Milky Way is but one among literal billions of such island universes in a dynamic expanding cosmos that blazed into being with a Big Bang 13.7 Ga ago. What preceded that seems unfathomable and any purpose utterly inconceivable.

A brief look at stellar nucleosynthesis is in order, the process of creating new atomic nuclei from pre-existing nucleons, primarily protons and neutrons in stars. These are the ongoing supply chains of the elements and involve complex reactions in very high-energy states. Every star is doing this.

The very first nuclei formed about three minutes after the Big Bang, a circumstance not anticipated by our Hellenic forebears and their element of the month club that worked so hard on expounding their cosmos. H is the simplest element on the periodic table and it is philosophically engaging that the universe appears to have entirely commenced thus. It is still 90% H, indicating that the cosmos is not that old in terms of enormous reserves for any projected future. We probably have not seen anything yet.

We think only H, He and probably some lithium (Li) and possibly trace beryllium (Be) were produced in the prime nucleosynthesis of the Big Bang and as for the current state of affairs we have discovered two sets of fusion reactions by which H is fused into He. With exceptions, the essential high temperatures and pressures generally only exist in stars. In those the size of the sun or smaller there is the dominant proton-proton chain reaction and in generally larger ones the CNO cycle. The latter is catalytic and occurs in stars greater than 1.3 M_{suns} and often those significantly more massive. Elements lighter than iron (Fe) are most often hurled into the interstellar medium when low mass stars eject their outer envelopes.

Successively heavier elements as far as Fe in the more massive stars are built up in the latter stages of stellar evolution by the triple-alpha process. Here two He nuclei (alpha particles) fuse to form Be which tends to swiftly disintegrate but upon being struck by another alpha particle produces a nucleus of carbon with six neutrons and six neutrons.

Depending on conditions He can fuse with C to produce O, He with O to make neon (Ne) and so on with progressively heavier nuclei. The heaviest elements of all are solely produced in the explosive nucleosynthesis in supernova explosions via the p-process, r-process and s-process (proton, rapid and slow). Supernova nucleosynthesis fusing C and O are thought responsible for the elements between magnesium (Mg) and nickel (Ni) and other heavier ones.

Gold, for example, was made in this special way and most of it naturally occurring on Earth long since sank to the core when it was still young and molten. The value we put on it seems quite arbitrary; it is too soft to employ directly. The crustal and mantle deposits were more likely delivered here by the impacters of the LHB. Diamond is an allotrope of carbon and has more practical uses, indeed the versatility of carbon is profound for what it can build.

We think that the stars creating carbon in the disc of the Milky Way were able to do so when they were red giants with three He-4 nuclei combine to make C-12, the most abundant isotope of carbon. Some of it seeps into the stellar atmospheres before being ejected as carbon enriched planetary nebulae such as The Ring Nebula. In the big picture it appears that carbon entered the disc of the Milky Way at a slower pace than oxygen.

Cosmic ray spallation is another natural process for the formation of chemical elements, the impact of cosmic rays on the interstellar medium. It is a significant source of lighter nuclei by the fragmentation of larger atomic species not directly produced in stellar nucleosynthesis.

Now on to the narrow "Goldilocks Zone" or "Circumstellar Habitable Zone" which is proposed as a limited distance from a star where water may exist as a fluid under a planet's appropriate conditions of temperature and atmospheric pressure. The postulated Goldilocks Principle invokes conditions just right in being not too hot and not too cold on planets not too big or too small. These are among other necessary factors of stability including protection from dangerous radiation. Speculating on the range of distances from the sun allowing liquid water may also be traced to *Principia* and the notion of a CHZ is first ascribed to Strughold along with the term "biosphere" about 1953.

About the same time Shapley introduced a "Liquid Water Belt" followed by Su-Shu Huang's assessments of a "Habitable Zone." It seems to work just fine here with all the right ingredients so we will start by looking at the solar system's other members.

The depth and chemical compositions of planetary atmospheres play the important supporting rôle as the eventuated diversities of Earth and Venus shows. Previously, its unknown surface conditions allowed optimistic imaginings but as information improved our fantasies proved to be fallacies. Our immediate neighbor after the moon is wholly unaccommodating to life and apart from size the days are gone when they were imagined to be twin planets.

It turns out that Venus possesses high atmospheric winds of a breezy 200 mph, sulfuric acid clouds dropping friendly H_2SO_4 precipitation and surface temperatures of a balmy 737 K. Due to the runaway greenhouse effect it is the hottest planet despite Mercury's greater proximity to the sun. The Cytherean atmosphere of dominant nitrogen is monstrous too, nearly 100 times the pressure at the earth's surface at ground levels.

The search for contemporary life on Mars proves negative but clearly we will never entirely give up the quest. It will remain an inexhaustible possibility that something organic existed there in the past. In search of real signs we could dig deep into the Martian crust at any location to chemically analyze the samples into literal dust. More pragmatically, the proposed Mars 2020 Rover should contain a conduction-cooled laser system capable of spectroscopically detecting carbon-based signatures of organic material and we look forward to *any* positive leads. Mercury and the moon have been written off as life supporting but the "Great Moon Hoax" of 1835 sold plenty of newspapers in New York.

Much can be said about potentially habitable or otherwise "Earthlike" planets and I direct the reader to the star Gliese 581 and its companions. It seems to harbor the champion of any known Goldilocks planet and at last there is a little real information on "superearths."

We are also fond of locating "hot Jupiters" and speculate on the dynamics of other solar systems with enough real data (barely) to make it acceptable science. This subject is also worth checking regularly but the caveat is that the ethos of Goldilocks Zones is an assumption. A CHZ may be temporal as well as spatial. We go along with it as a projection of biofriendly conditions in specific places and invoke the numbers game to project that there are millions of superearths out there. We also tend to disregard worlds of highly elliptical paths, binary systems or huge temperature ranges as suitable because they lack stability. This is more imagination than science but what can we do?

Accepting this, atmospheres and oceans can be highly protective and fully nurturing from hostile outer space and may not depend on such stringent locales. The pace of progress is generally very slow with distasteful mass extinctions always looming by changing conditions or outside interference like destructive asteroids impacts. Indirectly they can prove constructive by clearing the board for a new game to set itself up as with the rise of the mammals.

Perhaps all life grasps opportunity on the brink of oblivion. As a snapshot from evolutionary history we once had plenty of trilobites on Earth. Numerically, they were the most successful creature yet but they too disappeared in the Permian-Triassic mass extinction about 252 Ma ago. Such is their commonality as fossils down to our era that trilobite jewelry is affordably available. This particular Great Dying possibly happened in pulses and took out an estimated 96% of marine and 70 % terrestrial vertebrates over a very approximate 60 Ka. It is thought to have been the only mass extinction of insects to have taken place and things might have taken as long as 10 Ma to recover.

Expanding the speculation as we necessarily do, we postulate that some far bigger Goldilocks Zone exists as a Galactic Habitable Zone. We make correlations between a star's metallicity in broad types on the H-R diagram and the sort of planets they might have. It is the suggestion that different areas of the Galaxy are more biofriendly as whole zones in gross chemical attributes but there can be no proposal that stars spectroscopically like the sun or any stellar type are in some favored position to produce life-bearing planets. It is true that further from the galactic center the metallicity of stars increases, possessing elements other than H or He in generally greater amounts. Older generations of stars have less metallicity having been formed in the metal poor early universe, meaning that he cosmos is chemically evolving one star at a time.

Further from the galactic hub the harsh bombardments of radiation from neutron stars and X-Ray sources lowers and gravitational perturbations are presumably less disruptive. We reason that further from the crowded nucleus conditions may be broadly more amenable to life but in the outermost zones the heavy elements are less concentrated and the general level of star formation lower. This excludes the mass production of small rocky planets. I reject the suggestion that high mass stars propelling quickly through their life cycles do not allow sufficient time for advanced evolution on their planets or any chain of logic disallowing life. Are there rules of cosmic nature's thumb relating to biogenesis?

There is the further proposal that dusty spiral galaxies are more hospitable due to higher rates of star formation compared to elliptical galaxies composed of relatively older stars. Astrophysics shows this but astrobiology requires highly individual systems and protected ecospheres. It is productive to assess the cosmic abundancies of the elements but I suggest that a conceptualized GHZ might be pushing things too far with the present state of knowledge.

So back to hard data we go. The sun's path is approximately circular about the Galaxy, steering clear of the potentially dangerous spiral arms in its great path of orbit requiring 220 Ma, casually called a cosmic year. It has made about 20 orbits since its formation and maintains a speed of a calculated 143 miles/sec or 514,000 mph.

This may conveniently avoid the intense radiation and gravitational perturbations that on principle. We think the sun is traveling at the same rate as the rotation of the spiral arm and it remains up for grabs to establish what biological effects are caused thereby. Passing through denser clouds of material could be problematic and we have sketched the perils of supernovae, GRBs or marauding black holes. The thinness of the stellar population in our immediate locale might be a form of protection in itself.

The sun is typically metal rich for its type, further suggesting that it formed in a more metal abundant zone then migrated to its present position. Certainly, a metal rich interstellar cloud is more likely to possess rocky planets including the cores of potential gas giants. Calculations from Lineweaver *et al* place the GHZ as a ring 7–9 kiloparsecs in diameter to include 10% of the stars of the Milky Way and between 20 and 40 billion suns. More conservative estimates cut these figures by 50%. Unfortunately, it seems one of those areas where we will never know how right or wrong this might be.

One constructive challenge to the GHZ concept involves the older structures of the Galaxy, specifically the globular clusters that orbit the galactic core as satellites. They are compact and tightly bound by gravity having long since adopted spherical shapes for their high stellar densities. Assuming a few unseen members, approximately 180 of them exist in the Milky Way. Surely there has been far more time for anything including life to develop so who knows what heights civilization has risen to where the logistics and distances of real interstellar travel are less formidable? As ever, the possibilities are infinite.

The Origin Of The Solar System

There are no shortages of creation myths in cultural antiquity and the great religious traditions. Incidentally, only the Hindu tales touch on the right temporal ballparks with myths of the cosmos passing through cycles within cycles for eternity; meaning that the world's oldest religion may have grasped some element of truth for 5,000 years.

We will now construct an abbreviated set of highlights as it unfolded in Western science with the usual obligatory starting point in ancient Greece. Accordingly, the first prize goes to Aristarchus and his heliocentrism for indicating that any sort of planetary system exists and the earth is not central. It is regrettable that such a cosmography did not win out far earlier, either in the Hellenic world or among any of their intellectual descendants. Copernicanism was first made public in 1453 but was still formally heretical up to

1822, which beggars belief, given the scientific advances made by then. The term solar system only emerged in the early 18th century, a bit slow in coming.

Descartes proposed a model for the birth of the solar system in *Le Monde*, composed in 1632. It involved swirling celestial vortices, where particles from a large one condensed into planets that endure to this day in great tracts about the sun. Caution was still prudent in the propagation of scientific creation and it was only published after his death in 1664, the lesson learned from Copernicus. When Newtonian gravitation showed that matter does not behave in this way the vortices were banished forever.

The initial proposal of a nebular hypothesis is credited to Swedenborg in 1734, expanded on in principle by Kant in 1755. The former goes down more in the mystic than scientific category but it was a rare Kantian error of practical reason that the observed nebulae were planetary systems directly seen in development. It looked a promising idea but this is definitively not the correct interpretation, even as improved telescopes revealed far greater detail. Exactly what the spiral and other nebulae represent was debated up to the 1920s.

In 1770, Leclerc suggested an origin by a comet colliding with the sun sending great sprays of material into space, seeing fit to extend the timeframe of Earth history to 75,000 years. It was a step in the right temporal direction, but Pierre Simon de Laplace (1749–1827) refuted the ideas by proving that planets formed in this way would crash back into the sun.

The broad solar nebula hypothesis descends from the independent work of Laplace although it was more a casual suggestion than a firm proposal in 1796. The young sun was conjectured to have thrown off concentric rings condensing into the planets one by one as the gas contracted. A vast disk of material surrounding the hot center is tenable but it has been shown that dynamically, disks thrown off would not condense into discrete worlds. It brought on as much controversy among politicians and theologians as scientists of the day.

There is the story of his conversation with Napoleon in 1802. Having perused their great written works of science, the Emperor noted that Newton mentions God but Laplace does not even once. The scientist's alleged reply was that he had no need of that hypothesis. Better authenticated is his statement that the true object of the physical sciences is not to search for primary causes but to search for laws according to which the phenomenon are produced. Herschel also met Napoleon and later noted that he affected far more education than he really possessed. Is he history's noblest loser?

The Russian polymath Mikhail Lomonosov (1711–1765) had advocated hundreds of thousands of years for the causes of planetary formation including the earth. In the following decades, Lyall's Theory of Uniformitarianism put forward that the physical processes are effectively ongoing in geology. This was deduced from the relative aging of geological strata and opposed the ideas of catastrophism whereby the earth formed in abrupt and short-lived violent events. Lyall's *Principles of Geology* of the 1830s was influential in establishing that the present is the key to the past.

Uniformitarianism advocated that slower incremental changes gave rise to major geological features. James Hutton was first to suggest that the earth is no less than a super organism and notably concluded that there was no vestige of a beginning and no prospect of an end. Meanwhile, the Neptunists clung to the concept of a global flood with Noah and family bobbing about in his ark, accompanied by two of every creature for insurance against extinction.

By the 18th century discussing matters of the age of the earth had assumed a certain intellectual fashion as the required time scales expanded dramatically. Unfortunately, the whole nebular hypothesis fell out of favor, mostly due to the problem of angular momentum, which we will address. Too much exists in the orbits and spins of the planets and much too little in the sun.

As for the earth itself, that volcanism provided the first reserves of water, air and carbon compounds by huge outgassing was first proposed in the 19th century. Generally, Plutonism or volcanism was seen to better suit further geological findings. It showed that many rock formations are solidified from a molten state in addition to the sedimentary action of the sea. Fossil evidence served again and the story of our own world proved a great teacher about authentic time scales. Wells elucidated Earth history as a stupendous 2 million years in length, emphasizing that the forms of living creatures were anything but fixed and final. Plate tectonics, sea floor spreading and other advances like paleomagnetism and radiometric dating were to more than double the figure.

At the turn of the century Moulton showed that the nebular hypothesis ran into serious difficulties with the observed angular momentum of the system and collaborated with Chamberlin on a planetesimal scheme. Here, another star passed close enough to the young sun to cause material to be repeatedly ejected. In this scenario, spiral arms extended, partly falling back into the sun with other sections forming plantesimals that then cooled and coalesced into planets. The overall theory fell from favor but the concept of planetesimal accretion was retained.

James Jeans proposed a tidal theory or near collision hypothesis in 1917 in which a close approach of another star drew huge filaments of material to engineer the planets. Into the 20th century several such versions were put forward, mostly replacing the ideas of contracting discs. Jeffreys showed that a close encounter of a rogue star drawing great tendrils of matter from the sun was extremely unlikely. Russell pointed out that angular momentum alone would effectively drop them back into the sun and that material with sufficient speed to escape the immediate solar gravitation would then leave its environs entirely. We met this objection with the fission theory for the origin of the moon.

In 1939 Spitzer further concluded that any such column of material would more likely dissipate than condense. Also, the close passage of stars in this region of the Milky Way was now seen as statistically remote. Most such concepts were discarded by the 1940s including Lyttleton's scenario of the sun actually colliding with another star.

Until recent decades, we regarded planets as rare commodities and sought a special case to expound the origin of our own. In retrospect it was the absence of evidence — interpreted as evidence of absence — for the unseen presence of planets accompanying any other stars. They are far less massive and luminous than the smallest binary star companions observed and we blandly assumed they did not commonly exist. It is far more likely that planetary companions are difficult to detect than that the solar system is unique.

In 1943 Otto Schmidt proposed that the sun had passed through a vast interstellar cloud and became enveloped in gas and dust. There were variations on this theme, until we found that the time needed to form the planets from such a tenuous envelope of material far exceeds the approximated age of the solar system. A whole Russian school tended to suggest that the planets formed considerably later than the sun.

Carl von Weizsäcker developed more promising models of a solar nebula in the 1940s. It was more refined than the earlier ideas of Laplace, and Dirk ter Haar modified the Weizsäcker one further. This now included how the planets formed by accretion and Kuiper postulated how large gravitational instabilities could take place in the solar nebula promoting massive condensation. He saw the matter of the sun's slow rotation as an astrophysical problem in our understanding of G type stars. Whipple suggested a young sun of negligible angular momentum capturing a secondary cloud from which the planets derived but the idea gained little acceptance.

After musing on a nova then a supernova theory of origin, Hoyle also reintroduced a Laplacian scheme in the late 1950s with stricter mathematical detail. We began to visualize a proto planetary system contracting under gravity from a primeval great solar nebula as basically accurate. All evidence suggests that the sun and all its planets emerged in extended unison.

Further proposals from Urey and others place the primeval sun still surrounded by a nebula primarily consisting of H, He and dust particles where collision and friction formed a great disk like cloud in its equatorial plane. The innermost area underwent such pressure that thermonuclear reactions finally commenced and literally, a star was born deep within the great contracting nebula. We will proceed from this point as the accepted birth of the solar system and the very simplified order of events runs:

1. A primeval solar nebula begins to collapse under its own gravity.

2. Heating is most pronounced at the compressed center.

3. A protostar forms and also a gaseous accretion disk around it.

4. The star ignites but the disk radiates energy away and cools.

5. Dust particles collide and assume growth, eventually a runaway growth.

> 6. Protoplanets emerge in two broad groups i.e., small rocky and big gaseous types.

> 7. Modification and possibly orbital migrations take place.

The nebular hypothesis proposes a collapse from a giant molecular cloud and this is generally accepted now. The central protostar receives most of the material but due to rotation centrifugal force forms an accretion disk that radiates away energy and cools off.

By some estimates the primordial cloud measured 65 light years across with fragments less than 1 light year in size. This scale possibly gave rise to several new suns, which is consistent with other indications but these figures are highly tentative.

Let us take stock of some authentic situations. The outermost parts of the Oort cloud lies 100,000 AU from the sun, rendering the furthermost physical outposts of the solar system just under 600 light days from its center. Placing the heliopause at 123 AU from the sun, the Oort cloud approaches an impressive 2 light years and halfway to the nearest star. In many ways the local planetary system is larger than the furthermost orbits of the classical planets and Kuiper belt objects including Pluto. Simple calculations reveal that compared to 65 light years, the solar system eventuated at less than 3% of the scale of original molecular cloud whence it came.

This is a bit hazy but more reliable was its chemical composition. It was decisively mostly H and He in bulk with 2% heavier elements left by the nucleosynthesis of earlier stars and their prior ejection into the interstellar medium. Grounds to suppose this lie in the oldest inclusions within meteorites where traces are revealed of stable daughter nuclei of short-lived isotopes that probably formed in exploding, short lived stars. It is fortunate that so many meteorites are available for study and surely yet more information could be wrested from them.

Several supernovae may have occurred close to the sun whilst it was forming. The resulting shock waves (not gravitational waves) from them triggered denser regions of the pre solar nebula to collapse inward a little longer ago than 4.6 Ga. Shortly afterwards, the combined forces of gravity, gas pressure, magnetic fields and speeding rotation caused a flattened spinning protoplanetary disk to accrue. For quantification, one estimated diameter of the primordial cloud lowers it to 200 AU with the sun capturing 87% of the total material and Jupiter assuming 71% of the remainder.

The increases in temperatures and the generation of heat derive from gravitational contraction as a proto planetary nebula develops and the gas at the center is further heated by compression. With a core of some 10^6 K hydrogen burning commences, signifying the end of the protostar stage as the new star assumes equilibrium and what is termed zero age main sequence.

As ever, the radiative output of energy generally balances the inward pull of gravity as it settles down to busy stability. Meanwhile, the accretion disk condenses into rock, metals and at greater distances, ice and more frozen compounds.

We should look to strike a balance between the specifics of our own planetary system and the processes going on elsewhere. These are safe assumptions. With more than one example to work with we find a welcome respite from amorphous exobiology.

In our locale, the proto planets subsequently formed by more gradual accretion after the inner star had turned on.

In the closer and hotter zones the denser iron rich compounds precipitated into iron rich cores with silicate materials of lower densities then further swept up. Electrostatic charges played a considerable part in the formative processes, as did the temperature gradient that is still present. The particles collide and grow towards small boulders and at relatively quickly into plantesimals. At length they aggregate larger and larger as their masses assume increasing gravitational fields of their own and at further distance from the center a runaway accretion takes place. In our case this refers to the dominant masses of Jupiter and Saturn.

The outcome was two distinct types of planets, the terrestrial and gas giants. The first are small, rocky and close to the sun and the other big, gaseous and more distant. On the primeval solar nebula there must have been varying densities and compositions among its original inner, middle and outer make-up and again it is apparent that temperature played a major part. A specific band structure model of compositional differences and other factors derives from Alfvén work in 1954.

By several causes, physical evolution possibly moved faster or at least dif-ferently up close to the sun. With the terrestrial planets, cores, mantles and crusts emerged as the innermost worlds of Mercury, Venus, Earth and Mars. Further out formed the colder and much larger gas giants of Jupiter, Saturn, Uranus and Neptune.

Among the outer cooler zones and lower rates of motion, ammonia (NH_3), methane (CH_4) and water (H_2O) solidified with sufficient size for their gravitation to attract gaseous H and He to form their enormous atmo-spheres of the gas giants. Somewhere between the present orbits of Mars and Jupiter there was a significant ascendancy in size. Solid accretable matter was presumably more common in the icy depths even if it proceeded at a slower pace.

In their internal structures the gas giants probably harbor major zones of transition down to mantles and eventual cores over hundreds of miles of increasing depth. The dimensions are unknown but are surely distinct to the crusts or any specific hard ground beneath the comparatively thin atmo-spheres of the terrestrial planets. Crusts, mantles and cores definitively exist with Mercury, Venus, Earth, moon and Mars and among the small, rocky types Venus has the densest atmosphere by far. I suggest the gas and ice giants have no specific regoliths or abrupt surfaces.

Any detailed assumptions for the inner structure of the outer planets are precluded. Only the topmost layers of massive atmospheres are directly visible and Uranus appears obstinately bland. It lacks the belts, zones and disturbances so colorfully active with the others. We can only muse on the states of low temperatures and super pressures prevalent far beneath the visible cloud tops; another case of only scratching the surface, actually the uppermost section of colossal atmospheres.

The existence of deep dynamic cores is shown by several causes, notably the strengths of magnetic fields, particularly that of Jupiter. This is apparent in all planetary formation, even big asteroids are somewhat differentiated but all relative sizes or ratios of core to mantle to atmosphere remain unknown.

Here then are several mysteries to unravel in future planetary science. We can make direct investigations with seismic studies on the inner planets but it is challenging to drop a probe into the Jovian atmosphere and have it function and send data over any significant length of a dive. In conclusion, there are ranges of physical models for the inner structure of all the outer planets.

Having adopted the broad Laplace-Kant theory of origin we have made some progress. For example, there is evidence that the younger sun was less luminous by a factor of perhaps 30% and presumably was more variable during its early era.

This gives rise to the faint young sun paradox whereby estimates suggest that it gave insufficient heat at its distance to maintain liquid oceans on Earth at a time when they definitively existed. This problem is not fully resolved besides the major adjunct question of how the earth became so very watery. Proposed mechanisms include a major greenhouse effect trapping heat, a significant radiogenic component from inside the earth that has since lessened and/or far more tidal heating from the closer moon in the past. We have seen what else this may have triggered. It certainly stabilized the axial alignment of its primary body.

Exactly how and over what sort of a period did this whole formation take place? Dust grains originally aggregated by those electrostatic charges in small pieces and we visualize that the earliest plantesimals may have reached a few miles in girth in a mere 10,000 years of development. Modelling indicates that in material mass the earth may have essentially formed over a period as little as 40 Ma.

These figures are hazy but the solar system clearly had a violent and fiery early development including the LHB, generally ascribed to the gravity of Jupiter. The non-reworked surfaces of members such as the moon or Callisto are crater-saturated, the products of maximum age and minimal change. Meanwhile the asteroids weigh in as the best available collection of primeval bodies, too small to have undergone much change other than collision and some fragmentation over the full panoply of cosmogonic time.

The evolved planetesimals still hurtle about the sun in a uniform direction and remain in the same broad equatorial plane of their parent star. The anomaly of Pluto's high inclination is solved by finally recognizing it as the first discovered member of the Kuiper belt and not a classical planet. Case closed that it is dwarf planet and one of a myriad in type. Furthermore, the interaction of the very young sun with closer stars in the first 100 Ma of its existence may have produced the anomalous orbits still extant for the more remote bodies in its gravitational grasp.

Returning to planetary formation it was by increasing accumulation that they developed their own significant gravitational fields, further accelerating the processes towards structural independence. We must observe that 99.86% of the mass of the entire system is now bound up in the sun itself.

The runaway accretion was most pronounced in the case of Jupiter. It possesses more material than the rest of the non-solar planetary system put together and gravitationally is only second in effect to the sun. Possibly the primeval solar nebula was of initial higher density at that specific 5 AU from the sun or wherever the most massive planet formed.

Jupiter eventuated as a smaller solar system in itself with 4 large and a collective 63 known moons in attendance. For several reasons it seems likely that the Galilean satellites formed in its orbit. Somehow, Saturn is of low density despite its volume; see those cute drawings of it floating in a bathtub of water. When visiting Saturn see if you can find a date stamp and read if its rings are relatively old or young as formations. We have 4.6 Ga to play with but we cannot make up our minds about this particular feature's general age.

We therefore posit a relatively rapid accumulation of planetesimals and embryo planets as their gravities swept up more and more material from the proto-planetary nebula. The more remote worlds remain huge, cold gas or ice giants to this day showing that the outer accretion discs never underwent high exterior temperatures. The radial distribution of temperatures outward from the center reigns to this day and the stability of their orbits in addition to diverse gravitational resonances among them indicate situations long established. There are of course other sources of heat than direct solar radiation.

We have referred to the Nice model suggesting that the outer planets may not have initially formed where they presently lie and that they emigrated outward. One inference of the revised model includes a "jumping Jupiter" clause where another ice giant became ejected entirely out of the solar system and major migrations into the present orbits of Jupiter and Saturn.

For the Scattered Disc Objects, accretion would have been slower at those imposing planetary distances before the primeval solar nebula dispersed. Hence the dispersed disc of the present Kuiper belt originally developed considerably closer to the sun.

On the subject of asteroids, it was long discussed whether either accretion or fragmentation gave rise to the tens of thousands of minor planets still lying in a relatively narrow zone always referred to as the asteroid belt.

Shortly after they were first discovered in the early 19th century, Olbers suggested that they formed from a destructive disruption of a planet previously orbiting between Mars and Jupiter fragments still abide. We now think the asteroids were more a failure to develop into an individual small world due to Jupiter's gravity than the breakup of an existing one. In any event, their collective mass equals only 4% of the moon's. Low accretion is the preferred mechanism to make asteroids in the first place whilst fragmentation played a part in later stages. The uniformly primeval appearances of the visited asteroids, the recently confirmed differentiations of 2 Vesta and 1 Ceres and other factors denigrates any single lost planet.

Unlike Theia invoked in the making of the moon, the lost world of Phaeton never existed and as one paper put it, breaking up is hard to do. We now hold that the enormous space weathering, micro meteorite and greater bombardments leave them less than entirely pristine but the primeval nature of asteroids is important to the general story. Most conveniently for the studies of the very oldest material available, meteorites still drop in regularly.

Now, for that standing problem cropping up repeatedly: How did so much angular momentum become stored in the planets' orbits with a comparative low amount in the sun? (Mestell965.)

In physics, angular momentum is the measure of a rotating body to maintain rotation and it is always conserved. Note that the magnitude of the mystery hampered the very first enunciations of any proto nebular hypothesis.

By the present epoch, such a slowly spinning sun is not easily elucidated by a nebula spinning faster as it collapsed. We have established how the molecules within it collided with ever-greater frequency and that about the same time as their kinetic energy was converting into heat the proto planets accreted. This explains much for the system's origins but after 5 Ga the sun should still be rotating considerably faster than it demonstrably does. Hats off to Galileo for observing that the sun turns on its axis but the rotation period of 24.47 days at the solar equator does not fit the greater dynamics. Moreover, the sun continues to slow its rotation.

Theorizing a much more rapidly spinning sun in the past does not explain the facts of the case. Somehow the classical planets acquired 97% of its entire angular momentum with 60% of that in Jupiter. Equally, only 0.3% of the total now resides in the sun. Calculations show that if all of it held by the planets were to be transferred to the sun it would spin faster by a factor of 100. This is a stark contrast to the sun being 1000 times more massive than all the planets combined.

How can we expound that the lion's share of angular momentum still resides in the outer solar system? Additionally, the gas giants still rotate rapidly on their axes, as swiftly as ten hours with Jupiter and Saturn. The most massive pair are the fastest spinners, indicating that apart from their orbital motions some strong spinning up took place during development. Presumably they have slowed in the last few Ga but this is still a remarkable anomaly in the distribution of angular momentum. There is no case for tidal braking by the inner planets slowing the sun. They are too low in mass plus other recognizable effects like captured rotation would be apparent.

What are the possibilities? We pose serious differences and events in the deep past like intense coronal mass ejections transferring angular momentum to the protoplanets but this too is guesswork. A radical notion is that the solar system was far more massive and voluminous in its early era and that the accretion process was not that effective. Much matter not captured by the sun transported itself back into interstellar space, taking plenty of angular momentum with it. It is one proposal for the origin of the Oort cloud though I prefer that to be some surviving periphery of the primeval nebula. Its shape alone indicates this.

Perhaps the solar wind acted more like a gale long ago. The high velocity particles traced out the magnetic field lines and as the greater rotating field of the sun attempted to drag them around with it there it acted as a brake. It the fullness of time the vast numbers of escaping particles carried away most of its angular momentum and the solar spin retarded.

The consensus is that the sun gave away most of its angular momentum at an early stage by some combination of the actions of its gravitational and magnetic fields. Possibly, the latter were once significantly higher and its field lines within the primeval cloud resisted twisting, dispersing angular momentum from the center to the edges. The mystery is unresolved and remains a weak link in the accepted condensation theory. On this line of reasoning, the sun might possibly have taken a more active part in the formation of the inner planets and not merely served itself.

Instead of castoffs whose material was no incorporated into the young sun, the closer worlds might have been actively fed, leaving the outer member to their own devices of further evolution. We assume a sequential condensation and aggregation with the whole process tailing off as the supply of solid particles generally lowered. Accretion was strong up to a point then stabilized. In terms of the ongoing interaction of the earth's magnetosphere with the sun and the commanding effects of its energy and gravitational field we are still directly connected.

Let us be strongly optimistic about expounding all this in the future because there are endless locales and feasible projects for collecting samples to give us a better big picture. Should we reach a place where mysteries like the giant core of Mercury, the formation of the moon, and the slow retrograde spin of Venus, we could then reach into the very origin and history of the solar system. Of course there is the perennial question: Was there ever life on Mars?

A Quintessence Of Dust

Review the case of the Martian meteorite ALH 84001 where the *possible* vestiges of ancient microorganisms may have been identified. This could be highly significant — if we could only decide — but the remnants of bugs dead for billions of years on Mars is not exactly what we have in mind for discovering other life in the universe. The 4.3 pound achondrite meteorite was located at Allan Hills, Antarctica in 1984 and its concluded history includes crystallization from molten rock 4.091 Ga back in time, removal from the Martian surface by impact 17 Ma in the past and arriving on Earth 13,000 years ago. At the last count there are 132 recognized Mars meteorites on Earth but this is the only known specimen of specific age.

More general evidence surrounding the chondrite aging of meteorites places the dust sedimentation from the primordial gaseous discs about the sun 4.56 Ga in age. This is determined by the measurement of decay products of radioactive isotopes. We will briefly brush up about isotopes where two or more forms of the same element contain equal numbers of protons but a different quantity of neutrons in their nuclei. They therefore differ in relative atomic mass but are the same in chemical properties as the radioactive form of the given element.

A half-life is the time required for 50% of the parent to decompose into daughter isotopes and the natural clock most suitable for meteorites is the decay rate of rubidium (Rb) into strontium (Sr), which have a half-life of a full 49 Ga. As this is greater than the age of the cosmos so far analyzing the specific isotopes ^{87}Rb in its decay to ^{87}Sr makes astonishing precision achievable.

The dates obtained are consistent with the radiometric ages of the oldest known terrestrial and lunar samples and we now turn our attention to the oceans and atmosphere of the earth. We shall start simply enough. The oceans have been here a long time but we cannot easily account for their scale. Since the origin of the earth the atmosphere has changed greatly with biological processes playing a great part.

Until the earth cooled from molten form it probably had only a passing envelope of stable gases surrounding it, possibly it had virtually no atmosphere. The present one is at least secondary, hugely adjusted in composition from the original.

The first proper version was presumably captured and maintained by gravitation from the original solar nebula and may have been modified by the intense T Tauri winds of the younger sun or entirely stripped away. This probably took place with all the inner planets as their primordial H rich atmospheres tailed off into space long before any other compounds formed or the plethora of impacts commenced.

Such a cascade of radiation and particles could have emerged at approximately the same time as the accretion disc stopped feeding the young star. The wind is in collision with the inner zones and partly halts the growth process of the young sun. T Tauri stars are named after the prototype of a class of variable star in the process of contracting to the main sequence.

They are generally found near giant molecular clouds and roughly half of them possess circumstellar disks, genuinely thought to be the clumpy progenitors of planetary systems that we so energetically imagined. Such circumstellar disks probably dissipate on timescales like 10 Ma and possibly the active magnetic fields and strong stellar winds impart much angular momentum to the protoplanetary disk. (This would be nice for our theorizing. Is it a standard feature?)

The Earth hugely adjusted its atmosphere by the processes of hot outgassing from within. As the mantle solidified and much of the metallic iron sank to the core, oxidizing conditions were promoted in the relatively oxidized mantle.

This produced gaseous nitrogen, sulfur and carbon and perhaps 200 times as much CO_2 as is now present. Its secondary nature is confirmed in the relatively low abundances of the noble gases neon, krypton (Kr) and xenon (Xe) on Earth compared with their occurrence in the sun.

As for gross physical structure, the young Earth had been heated to a nearly fluid state by radioactive decay as denser matter was able to fall to constitute the deeper and core regions. An upper differentiation developed into a mantle and upper crust as oceans of magma cooled and solidified. Metals sank to the core with an overlying mantle formed by molten rock. Degassing specifically occurred during the later growth stages when high temperatures drove gases from the solid material. Further and more general modifications of the atmosphere resulted from those biological cycles and processes over time, which are notably special features.

In addition to major plate tectonics and sea floor spreading our world is anomalously watery. We have comparatively reliable ideas about the evolution of the atmosphere but how did the earth acquire so much water? How plausible is it that the earth primordially formed with the bulk of its water already present as chemically bound up? It was probably ½ Ga before the earth's surface was both cool and solid enough to allow the collection of water on the surface.

There is far too much liquid water here than can be easily explained. The question is more fundamental than the observation that the earth's surface temperatures and pressures allow it to exist as a fluid or ice. Studies including those of the mineral zircon indicate that liquid water existed in bulk as far back as 4.4 Ga.

So much water may have a partial explanation of origin. Some off it may be attributed to chemical and material loads imported in cometary sources, there being enough H_2O tied up in the frozen hydrocarbons and ices comprising comets to partly support the contention. The slow leakage of hydrated materials may have been another contribution to the enormous amounts of ocean, lakes, rivers and groundwater we see. Major volcanism where massive water vapor was placed in the atmosphere is another possibility, later condensing to rain and standing liquid reserves that very early in the story.

Such enormous outgassing could be a viable mechanism for producing both the air and sea. Prodigious amounts of steam may have condensed into water as the earth cooled with salinity deriving from chemical interactions with the crust.

Hypothetically, the collisional ejection forming the moon may have left a rock vapor atmosphere condensing relatively swiftly into a heavy CO_2 blanket, containing sufficient water vapor to massively reacted with prevalent hydrogen. Despite an estimated temperature of 230º C the atmospheric pressure of the CO_2 allowed the oceans to stabilize and not boil away.

The broad hypothesis of huge quantities of terrestrial water supplied by thousands of cometary messengers is supported by the isotopic ratios of hydrogen to deuterium in seawater. We are confident it played some part. Certainly the water, atmospheric gases and carbon compounds formed the vital precursors to the rise of elementary life and prebiotic molecules may well have first sunk into the oceanic depths as opposed to welling up from the lower crust.

In either case the conditions there remained far more stable in temperature and protected from solar radiation. The general term Hadean (after the Greek *Hades* for underworld) applies to the periods prior to 3.8 to 4.5 Ga in the past as the earth settled down to individuality. We saw by the process of collisional ejection that the moon soon joined it and definitively was an early event of importance to the outcome.

Over time, modification of the atmosphere by biological cycles and photosynthesis by green plants and algae played a major part. Welcome to the Great Oxidation Event, which occurred at a slow pace even by standards although its exact chronological place is not precise. Also known as the Oxidation Revolution it was a major environmental change in the biologically induced appearance of dioxygen (O_2) in the global atmosphere.

In action, cyanobacteria's oxygenic photosynthesis might have taken from 3 to 1 Ga to generally oxygenate as the first microbes capable of producing oxygen by photosynthesis or certainly on such a scale.

Previous to the GOE any free O they produced was chemically captured by dissolved iron or organic matter at up to this point these O sinks had not been saturated. Major geological evidence for this includes the huge deposits of insoluble iron oxides appearing as banded rock laid down from the Archaean and Proterozoic eras.

The GOE was the point in time when the minerals became saturated and were unable to capture any further O. The excess of free O began to accumulate in the atmosphere and being toxic to obligate anaerobic organisms the rising concentration probably destroyed most such organisms.

Another long term effect of so much free oxygen was the removal of methane as a greenhouse gas and the cause of the Huronian glaciation and possibly the longest and most severe ice age in Earth History dated approximately 2.4 to 2.1 Ga ago. A repeating cycle may have occurred in causing major glaciations. Cyanobacteria in action during the warm period produce O, uses up CO_2 and removes methane (CH_4) causing temperatures to drop. This slows down the bacteria and the temperatures rise again.

Alternatively, it is also possible that there was a 250 Ma slowdown of volcanic activity reducing the CO_2 levels and greater greenhouse effect. The Earth maintained itself above freezing due to the presence of methane and in it absence the temperatures temporarily dropped.

The present composition of the atmosphere has only existed for an approximate 850 Ma. It finalized as a mixture of 78% nitrogen, 21% oxygen and 0.9% argon with trace gases that are in descending order: Carbon dioxide, neon, helium, methane, nitrous oxides and ozone with tiny components of hydrogen and krypton. It is divided into five principal layers with the troposphere and stratosphere at the lowest reaches.

As always, the sheer activism of Earth sets it aside in the past, present and future. The only world with major ongoing geophysical activity is the special case of the Jovian moon Io. There, volcanism dominates to the extent of ever shifting surface features and volcanoes capable of lifting sulfurous material off the satellite. The discovery of these volcanoes was a significant step with the thin torus ring deposited in orbit previously detected from

Earth based observation. Hence, such a specialized form of volcanism had actually been suggested prior to the Voyager flybys. Gravitational interaction with Jupiter is the energy source for a small world endlessly squeezed in its passage and a submarine ocean in the case of Europa could derive heat energy by the same mechanism.

With Venus the atmospheric constituents are 96.5% CO_2 and 3.5% N with traces of other compounds at much higher surface pressures than Earth. Possibly we did have oceans in common as far back as 4 Ga. The sun's growing luminosity has long since driven a dynamic greenhouse effect that in turn produced copious amounts of CO_2 from carbonates of rock.

This raised the global temperature to the existing levels and long since eradicated any seas or lakes. It might expound the density of the atmosphere of Venus. On principle it would be straightforward to place a probe or rover on the Cytherian surface and determine if such features as oceans once existed.

Contemporary Mars is of the same broad atmospheric composition at 95.32% CO_2 and 2.7% N but at far lower pressure, a mere 0.7% of the terrestrial and cooler general temperatures. Conditions there are both dusty and variable with the Martian seasons and the varying sizes of the polar caps are a favorite for moderate aperture telescopes. Temperatures rarely reach above freezing and can fall as low as -89º C. Some meteorological and localized weather effects take place amidst the slower paced seasons and the Martian day of 24 hours 37 minutes is charmingly called a sol. On occasions, dust storms can engulf most of a hemisphere.

Mars probably lost the bulk of its atmosphere over time making conditions far drier and colder. We are definitely better informed about the Martian scenario including very small amounts of transient liquid water on the surface. No large standing bodies exist today under the contemporary conditions of temperature and atmospheric pressure although the appearance of dried up beds and former paths of stream are quite evident. Water left good geological evidence of itself and perhaps much more.

The entire record of the rocks might be more complete for a globe unaffected by massive plate tectonics or the upheavals common in terrestrial geophysics. One day we will have a clearer picture of how alike or diverse the primal conditions of Venus, Earth and Mars were and how they diverged as environments. In the future of manned space exploration the logistics and circumstances ordain Mars as the next port of call.

One piece of laboratory study that became one of the landmarks of modern science set out to address the earliest conditions of Earth. In the famous Miller-Urey experiment of 1953, the hypothetical conditions were simulated to explore the occurrence of life's chemical origins. It was to test the proposals from Oparin and Haldane that the primeval soup favored chemical reactions synthesizing organic compounds from inorganic precursors.

The experiment itself utilized water, methane, ammonia and hydrogen within sterile tubes with one flask 50% full of liquid water and another equipped with a pair of electrodes. Evaporation was caused in the water by heating the liquid water with sparks activated between the electrodes in the simulation of the action of lightning through cloud cover.

The sample atmosphere was then cooled so the water could condense and trickle back into the first flask, setting up a cycle. Within 24 hours the mixture assumed a pink color and by two weeks of operation it was observed that 10–15% of the system's carbon had become organic compounds. About 2% of the carbon went into the formation of amino acids that make up the proteins in living cells, with glycine proving the most abundant over cytosine, adenine and thymine.

In retrospect, forming amino acids in a laboratory is comparatively straightforward, whilst it is entirely beyond reach to synthesize DNA. Any artificial creation of life will remain in Dr Frankenstein lab indefinitely. The moral of that story is that he meant well with his cutting edge scientific curiosity but the experiments go out of control. It was written in the days when life was regarded as some magic event fired by some vital spark or mysterious force. Compare this to the success of the Human Genome Project, which has plumbed the complete, set of nucleic acid sequences for humans, encoded as DNA within the 23 chromosome pairs in our cell nuclei.

Research since has suggested that the primordial soup of an atmosphere might not have been as assumed but other prebiotic experiments under varying conditions do produce racemic mixtures of compounds of varying complexities.

Further, in 2007 the examination of the sealed vials from the first experiment subsequently found more than the 20 different amino acids reported by Miller. We often speak of the "building blocks of life" but amino acids are analogous to individual bricks and we seek to explain the design of whole structured organic buildings.

In the greater natural experiment of the early environment, life first emerged 0.5 to 0.7 Ga after the fiery origin of the world. Bacteria in their groupings may have been dominant for a full 3 Ga. Eubacteria (the true bacteria) were mostly photosynthetic using solar energy heat and light as their sole source of energy. They gave rise to eukaryotic cells containing nuclei and membrane bound organelles protecting their chromosomes constructed of DNA and proteins and slowly but surely more oxygen was provided into the atmosphere.

The less sophisticated and smaller prokaryotes have a single DNA strand floating freely in the cell. RNA molecules are similar to DNA but have a single strand rather than a double helix and acts as a messenger to the site of protein formation inside a cell where the specific RNA gives the information to synthesize a particular protein. The RNA strand matches with the DNA's base pairs and the process called translation encodes information about the new protein to be built.

The first photosynthetic organisms probably used hydrogen or hydrogen sulfide rather than water as the sources of electrons before the appearance of cyanobacteria. Photosynthesis is the process that converts carbon dioxide into organic compounds, especially sugars using the solar energy releasing oxygen as a byproduct. They were probably a form of algae that first efficiently used the energy from the sun. We think the chloroplasts of modern plants are the products of these ancient symbiotic cyanobacteria. Certainly, the Great Oxidation allowed for protection on the land surface of the Erath from the ultraviolet light from the sun and we deduce that life could not leave the seas until there was a protective layer of ozone.

Let us take a purely chemical approach. An organic compound is any of a large class of chemical compounds in which one or more atoms of carbon are covalently linked to atoms of other elements, generally hydrogen, oxygen or nitrogen.

The difference between organic and inorganic is the very presence of a carbon atom, meaning that organic compounds contain a carbon atom and often a hydrogen atom to form hydrocarbons whilst almost all inorganic compounds possess neither. The few compounds containing carbon not classified as organic include carbides, carbonates and cyanides. In chemistry organic basically means that a molecule has a carbon backbone.

* * *

What then is everything made of? Big Bang cosmology indicates primordial H and He but what is all this matter as it finally descends to us?

From Democritus to Leibnitz there is a trail of probing of what the world is made of and how it fits together. We can do so much better now that the applied philosophizing of our forebears and once again the intellectual and semi-mystical thinking of the past finally presented us with a more coherent picture.

Spectroscopy has resolved that the ten commonest elements in the Milky Way are hydrogen and helium as a preponderance followed by oxygen, carbon, neon, iron, nitrogen, silicon, magnesium and sulfur. Strictly, there are 88 regularly occurring natural elements on Earth and a few fleetingly brief artificial ones by advanced experimentation.

In detail, the elements 43 Tc, 61 Pm, 85 At and 87 Fr have no stable isotopes or long half-life so they are not considered naturally present; only small amounts are temporarily made in nuclear reactions with the latter two being the rarest naturally occurring element. If they are included, then so should be 93 Np and 94 Pu, which are produced by the absorption of neutrons in the spontaneous fission of element 92 uranium and 90 thorium — making a grand total of 94. For the suggestion that 100 makes a rounder number and arithmetic elegance we remind ourselves than the whole base 10 derives from the number of digits on human hands. 100 is of no greater significance than a tool of mathematical modeling.

The periodic table sets a major improvement over the philosophical earth, air, water, fire and special celestial stuff making up the "elements." They were never serviceable at coherently explaining the makeup of the land, air and oceans but in simplified meteorology the descriptive phrase stuck. We are at the mercy of the elements *etc.*

By real chemical definition we find none of them are elements at all. Rock and soil are diverse multiple compounds, water is a simple oxygen-hydrogen compound and fire is a powerful oxidizing reaction. See how the milestones of methodical discovery dispel the millstones of static thought.

Returning to nucleosynthesis, it was enthralling to read recently that the origins of solar system elements may now be confidently associated with its widely different forms. These are Big Bang fusion, dying low mass stars. Cosmic ray fusion, exploding massive stars, exploding white dwarfs and possibly, merging neutron stars. (Jofré and Das.)

We conclude the matters of bio friendliness and the whole subject of biogenesis with a simplified quote from the engaging book *Rare Earth* and its chapter on life's first appearance on Earth. The chemical evolution of life entails four steps:

> 1. The synthesis and accumulation of small organic molecules such as amino acids and molecules called nucleotides.
>
> 2. The joining of these small molecules into larger ones such as proteins and nucleic acids.
>
> 3. The aggregation pf the proteins and nucleic acids into droplets that took on chemical characteristics different to the surrounding environment.
>
> 4. The replicating of the larger complex molecules and the establishment of heredity. The DNA molecule can accomplish both but it needs help from other molecules such as RNA.

How Long Have We Been Wondering: Is Anyone Out There?

Some insightful deliberations on the idea of life forms existing beyond Earth are traceable among the esteemed commentators of the past. For example, the Greek Atomists Leucippus and Democritus held that multiple indestructible atoms could form other worlds and in a surviving work, the Roman Lucretius gave poetic eloquence about the movements and interactions of particles in empty space. On life beyond this world, the Pythagoreans asserted that the moon was home to creatures superior to us as early as the 6th century BCE and Plutarch (1st century AD) similarly contemplated lunar inhabitants. Epicurus in a letter to Herodotus deduced that there were infinite worlds and decisively, other creatures. The largely forgotten Chinese philosopher Teng Mu held that beyond our earth and sky there must be many others.

Kepler mused on a populated moon, rationalizing that if other globes are not meant for man's sake, how can we be the masters of God's handiwork? Just as the moon exists for us on Earth, Jupiter's moons were there to serve the Jovians. His posthumously published *Somnium* (1634) describing a semi-magical visit to the moon has been called the original SF novel, prior to the works of Poe or Mary Shelley.

Huygens wrote an imaginative piece concerning creatures on other planets and the concept appealed to William Herschel to the extent of ruminations on beings inhabiting the cooler layers of the sun. Kant himself considered the matter, the plurality of worlds becoming a part of intellectual conjecture. Once we had better recognized the nature of the planets they became potential abodes. Imagination is a powerful thing and anyone complacent enough to think that where we are in science is as far as it will ever go misses the essential point.

We have reasoned that the phenomenon of life lies somewhere between a unique fluke of this planet and a cosmic feature on a consequential scale. Either way it is a significant component of a mature Weltanschauung, so let us introduce the Principle of Plenitude, holding that what is possible in nature tends to become realized. It asserts that the universe contains all possible forms of existence. According to Lovejoy, this idea can be traced back to Plato's *Timaeus*.

Speculatively, we now shift gears from prebiotic chemistry into ultra long-range communications.

Taking it from the top, we are either alone or we are not alone. The calculated possibilities indicate a somewhat populated galaxy, but what sort of a quantifiable somewhat? Is it lonely or crowded out there?

The most conservative numbers plugged into the Drake Equation produces imposing quantities of living planets and it make a good parlor game to put in different values, turn the arithmetic handle and convince ourselves just how inhabited the Milky Way is. The recent and ongoing discoveries of exoplanets shorten the odds further.

Even the best-case scenarios for forging lines of communication have light years to contend with because the time required to send/receive anything cannot be circumvented. It is one of nature's unavoidable encumbrances that any exchanges will take at least decades. If we should ever rise to conversing across whole light years it would be an object lesson in patience. There can never be snappy conversations with even the most obliging neighbors although the time frames so restrictive to us may not be essentially so to them. We reiterate that the speed of light constrains the transfer of waves and any information. We do not know where to listen in with either frequencies or direction and conjecture still surrounds how to go about it.

There are volumes of engaging material on CETI but the feasibility studies are analogous to hunting quiet needles shrouded in haystacks of vast natural background radio noise. We have developed a few small magnets for the purpose but it wholly assumes that anything detectable is genuinely there in the first place. Listening operations from the original Project Ozma in 1960 up to the prospective Allen Array have not produced results so far. When we first commenced the enterprise of sending out intentional signals there was no hard proof that extra solar planets even existed.

This included the radio message aimed at M13 in 1974, which was a piece of functional art in itself describing basic information on Earth and humanity. Look up the "Arecibo Message" and see how the design depicts the atomic numbers of common elements, the formulas for sugars and bases, a graphic of DNA's double helix and more basic stuff like a human figure with a diagram of the solar system indicating planet of origin.

Technically, the message consisted of 1679 binary digits and approximately 210 bytes transmitted at frequency of 2,380 MHz and modulated by shifting the frequency by 10 MHz with a power of 450 kW over 3 minutes. The globular cluster M13 is 23,000 light years away. Since this there have been numerous deliberate messages sent out and the story so far includes sending "Across the Universe" by the Beatles to the star Polaris.

An indisputably artificial signal from the orbit of a close star would be ideal but things are emphatically not that easy. Obviously, locating the source is crucial but it could indicate that communicating civilizations are not close to us. The suggestion of the natural frequency of cold hydrogen gas at 1,420 MHz has been put forward as a common ground for communication but what patterns and modulations of a carrier wave might others be utilizing for the purposes?

Assuming we could achieve a breakthrough and establish transmissions, language should be a soluble problem although it would require serious work. Linguistic studies assure us that the matter could be addressed. With scientific facts and notations held in common and have already employed this with outgoing radio messages and the plaques on the Pioneers and Voyagers. Mathematics and functional chemical formulas describe patterns universally held in common and transcend the barriers of subjective language. It could therefore set the basis of introductory discourse.

We again limit the speculation to material cellular conscious beings evolved on the solid surfaces of planetary bodies but as always, who knows what forms intelligence has taken and how else they might communicate by laser or special applications of the electromagnetic spectrum? They could interact by means wholly beyond our comprehension.

From Marconi up wireless telegraphy is a recent addition to our technological toolbox and simple deduction shows that communicating aliens have means at least equal to ours. One tantalizing prospect would be acquiring not just a greeting but some massive amount of useful data from extraterrestrials; something outside our realm and above the previous human level of cognition.

Admittedly, it is overheated optimism that we could start juggling giga bytes of digestible material and images in a meaningful to and fro. How difficult is accidental eavesdropping compared to receiving generic addresses? Could we ever find a specific message directed at us?

Assuming so, I fear there is little of importance we could convey that they do not already grasp on principle. A catalog of images from planet Earth might be not be profound new knowledge from their standpoint, only detail and a novel example of a young technological civilization near Sirius. We are so backward that we previously knew of no other life in the cosmos, clearly a humbling state of affairs.

Take the notion that some other beings are fragmentarily aware of our existence. Could they have the technical sensitivity to access useful images if not translatable sounds at frequencies meaningful to them? Has anyone out there cracked human language? From our contemporary past, manufactured electromagnetic waves escape the ionosphere to stream outward from Earth, admittedly becoming greatly attenuated over distance and quite diffuse having crossed whole light years.

On the other hand, we have radio telescopes capable of communicating with its equivalent anywhere in the Milky Way. Whilst we have dispatched crafted messages from time to time, it is the movies, soap operas, documentaries *et al* that are our ambassadors to the stars. Here is the unintended outcome of FM radio and TV that rarely draws comment.

It seems inevitable that someone will eventually receive some dampened down whisper from humanity. It may not be their first indication of other intelligence and we might not be considered important. Who needs to converse with inferior jabbering monkeys hundreds of light years away? CETI is deductively sound as a raw possibility but there is the nagging possibility that we are going about it in entirely the wrong way.

We could hold that extraneous transmissions, generic addresses and willful attempts to communicate are already arriving here if we could only decipher them. If alien signals or voices are being transmitted, how would we recognize them? I am delighted to read that ceti@home is patiently combing for specific signals amidst great blocks of recorded random noise and has three million computer users. There are at least two other similar projects on the go, and if the group calling themselves "The Knights Who Say Ni" pulled it off, I would die laughing. In the earliest published paper to be taken seriously, Cocconi and Morrison pointed out that whilst the probability of success in interstellar communication is low, if we do not search then it is zero. Institutions like Arizona State University's Beyond Center, The Seti Institute and Breakthrough Initiatives and others take cerebral long-term approaches and deserve success.

Man has announced his presence and our existence is neither a secret nor can the announcement be retracted anymore. When radio and TV commenced, no one thought of spreading a spherical wave front into nearby space, currently over 80 light years in radius. We pump plenty of commotion that is not absorbed by the ionosphere and its source must be easily discernible by oscillating from one side of the sun to the other over a mere six months. Observationally, it would therefore be simple on principle to derive the period of origin of the artificial noise as we orbit. Of course we are already in space and on view.

This basic feature is recognizable even if the waves could not be more fully interpreted or deciphered. Moreover, the signal strength and continued to rise since the 1930s although this may have dampened down to more of a plateau with our more digitalized and optic cable communications of late. How much would be clearly recognizable is an open question. So what is our literal sphere of influence so far? Who or rather whereabouts are we potentially addressing?

Taking another look at the stellar Neighborhood there are 56 stellar systems within 5 parsecs of the sun. They are composed of 11 binary and 5 multiple star systems, the rest being single bodies. Fully, this comprises 60 hydrogen-fusing stars of which 50 are red dwarfs, 13 brown dwarfs and 4 white dwarfs incorporated over the same 16,000 cubic light years. The Neighborhood contains 3 naked eye objects, Alpha Centauri, Sirius and Procyon, which are familiar sights. Looking further, we approximate 15,000 stars to lie within 100 light years and with so many undiscovered due to low luminosity this is clearly a lower limit. It stands that this many independent star systems are already in range of communication and the wave front continues to expand.

N.B. In stellar astronomy it is a fallacy to suppose that only the immediate area contains so many small stars and that the greater neighborhood is specially populated with stellar glowworms. We simply are able to study the immediate locale better and locate them. There are only those three bright stars within that 5 parsecs and a mere 12% of the others in the vicinity may be seen without optical aid.

The 5 stellar bodies immediately beyond Alpha Centauri in random distance require at least binoculars to be observed. A huge commonality of low luminosity stars is clearly inferred from their distribution throughout this micro region of the Galaxy, which is no special location.

Furthermore, the object labeled WISE 0855-0714 lying at 7.3 light years appears such a sub of a brown dwarf it could be a rogue planet and only the second candidate of its class. (Read up on Cha 110913-77344 at as estimated 163 light years for the first evidence of such an orphan planet.)

Futurism And The Cosmic Facebook

The natural sources we study in radio and optical astronomy pack considerably more power than the artificial signals we seek in vain. Radio galaxies and pulsars are incomparably more forceful than ET's muffled chat if it even exists, whatever his intentions and entreaties.

Some more general and strong beacon of clearly artificial construction could exist in deep space, periodically or permanently signaling into the void, but no such entity of deliberate engineering has ever been located. Beyond temporary misinterpretations or severely wishful thinking, there seems not a hint.

We must not capitulate and assume that hard data about extraterrestrials lies forever beyond reach. The intelligent life discussion does not rest in metaphysical domains or such hefty speculations that to all practical intents and purposes, nothing can ever be substantiated.

I will always advocate that unknown does not equal unknowable. CETI has occasionally drawn criticism where we would be better off trying to contact the dead or studying ghostly goings on. It has been vilified as a pointless or even dangerous pursuit but the unequivocal discovery of intelligent life in the universe remains one of the holy grails of astronomy. It is not waste of resources and it is probably best that the enthusiasts are not government operations.

In practical astronomy, star atlases informatively give the positions, luminosities, distances and spectral types and whole character of stars. Much useful data has been painstakingly researched and confidently listed in their pages as tools and guides to the firmament of the night sky. Now here is something adding a new dimension to hands on observation, communing with the cosmos or however we term the noble pursuit of scientific stargazing.

For example, when I use a telescope to star hop to an asterism known as the Coathanger, I take pride in knowing the way from Altair to Sagitta, past 9 Vulpeculae and a few degrees northward further to reach the same Brocchi's or Al-sufi's cluster. In practical astronomy we tread beautiful celestial paths to locate the Andromeda galaxy or the Messier object of choice. Call me old fashioned but using Goto technology to locate objects in telescopes can take some of the fun out of it, efficient as it is. The vital point is that charts and eyeball observations often must present someone's home star. Should we ever be able to take the momentous step forward to know, it must involve a known star or cluster. In a way, the solution is right under our noses.

Dear Charles Darwin

Around distant stars are many planets harboring idyllic tropical lagoons of planetary seas with a protective atmosphere. There, biochemical processes are commencing that will culminate in life. Time, as they say, is of the essence. Let's let Darwin do the talking for a while — here's the conclusion of *Origin Of Species*.

> There is a grandeur in this view of life, with its several powers, having been originally breathed into a few forms or into one: and that, whilst the planet has gone according to the fixed law of gravity, from so simple a beginning endless forms more beautiful and most wondrous have been, and are being, evolved.

Note that the final word was the only use of "evolved" in the entire work.

The matter of "Darwin's delay," whereby the written work was ready to go for two decades before it finally saw publication, was due to his severe reservations for upsetting the apple cart of established belief all over again. Darwin himself disliked controversy and a sort of intellectual timidity delayed the full release of his researches. Yet here for the first time emerged the serious proposal that humans resulted from a monkey ancestry. Furthermore, all organisms share a common ancestor.

This was a bolder and more controversial stroke than Copernicus had unleashed in decentralizing the world because it engaged us as beings, not merely the physical stage of our existence. Darwin once wrote to his friend Joseph Hooker and hit the speculative nail right on the primordial head:

> But if (and oh what a big if) we could conceive in some little pond, with all sorts of ammonia and phosphoric salts, lights, heat, electricity etc., present that a protein compound was chemically formed ready to undergo still more complex changes.

This is remarkable scientific prescience. Both Copernicus and Darwin strongly bucked the traditional and prevalent views of their separate societies. Philosophical adjustment to heliocentrism had been made by Darwin's era yet he is criticized and disparaged to this day.

There are many thinkers and scientists we would love to bring up to speed in our era. What would Babbage and Pascal think of personal computers, or Kepler of the mission that bears his name? What would Newton have to say about the detection of gravitational waves? Finally, what do we have to tell Darwin today?

Well Charles, first we'll update you about passing on the basic characteristics of living things for the perpetuation of a given species. These are the nuts and bolts of natural selection and your survival of the fittest. Deep down where it counts we all have a lot in common, possibly a single original ancestor.

The directions for building a cell are recorded in DNA, which is faithfully copied and transmitted to each generation. DNA is therefore the long-term storage of information like blueprints conveying the instructions required to build the components of a cell.

The segments of DNA carrying the data are termed genes and within cells, it organizes itself into lengthy structures called chromosomes. DNA consists of two polymers made of simple nucleotide units, their backbones built of phosphate groups and sugars joined by ester bonds. A "negative" copy of itself carefully protects the stand and these two complementary strands are wound together in the form of a double helix.

The resulting twisted coils are wound upon themselves for compactness. Attached to each sugar is one of four molecule types called bases. These are adenine, cystocine, guanine and thymine abbreviated to A, C, G and T.

The molecular strands of DNA are of considerable length (about a collective meter in a single human cell) and are located in the cell nucleus of higher cells and in the cytoplasm of bacteria. Fragmented into 24 pairs of chromosomes the DNA holds the recipes to make thousands of proteins of different types and of diverse function. The code is read by copying stretches of DNA into the related acid RNA by the process of transcription. The largest human chromosome is about 220 million base pairs long. The whole DNA chain is 2.2–2.6 Angstrom units wide (2.2–2.6 nanometers) and one nucleotide unit measures 3.3 A or 1.33 nm long. Even the smallest known virus comprises a loop containing some 10,000 nucleotides, hinting how organisms are able to self-organize and reproduce successfully.

Now Dr. Darwin, have you caught up about the commonality of planetary systems? There were previously hard to detect and we made the premature judgment that they were uncommon. Do read up on the Big Bang and that expanding universe stuff too. It is actually accelerating.

We safely assume that there are likely lots of busy microbes inhabiting diverse planets but there are comparatively far fewer animals and even fewer cerebral creatures bothering their heads about it. We have not really cracked biogenesis or the existence of any other life in the universe but thanks for getting the ball rolling. You did for the life sciences what Copernicus did for astronomy.

We think in terms of an extended part played by microorganisms and elementary plant life shaping the greater environment, principally the earth's atmosphere. This is acknowledged as a slow paving of the way to the complexities of highly specialized cells. Further development seems leisurely in the extreme with fits and starts over enormous time. Everything was limited to bacteria and other unicellular organisms for fully 3 Ga. Yes, the world is really that old. The geologists of our day first got an inkling of this scale of time just as the astronomers were discovering the true dimensions of space.

Yes, the world is really that old, leaving plenty of rocks and fossils to interpret. In context, non-microscopic small animals and plants began in the sea a mere 700 Ma ago, emerging onto the land 400 Ma in the past. Between these eras there was a veritable Cambrian Explosion. We estimate multicellular marine plants appearing some 600 Ma, with land plants beginning to dominate the continents a mere 420 Ma before the present time. In a way, they still do — considering their biomass and the importance of the carbon cycle. The whole diversification of life still defies our understanding but we've got a lot right. The range of biomes is so dynamic that some creatures thrive in environments that would instantly kill others, air and water being the principal examples.

One proposal for life's origin is a great rain of prebiotic chemicals permeating down through the atmosphere with dust enclosed in tiny foam bubbles. Or reaching down to a clay bottom at the bottom of the sea could suggest how the first cell walls formed. This could have been the critical step, if we use Occam's Razor. It certainly has some appealing points.

By means better understood now, the most primitive organisms self-organized and learned to replicate reliably. Amino acids acted to form short chains of peptides whilst nucleic acids were able to join them in catalytic loops. Biochemical reactions links DNA fragments to peptides and organisms learned to multiply by the encoding of amino acids. They are in relation to an external environment yet aspire to be independent of it. Life is a runaway success and occurs in a myriad places in the cosmos. Our final word we find in Richard Fortey's *Life*:

> Some writers, especially geneticists, invoke the imperative drive for reproductive

success, the urgent bidding of DNA to propagate itself until it subdues the very

stars.

Epilogue Of Quotations

Ed Mitchell was the Apollo 14 astronaut and sixth man on the moon in February 1971. Having something of an epiphany on the return voyage he said the following:

> Instead of an intellectual search, there was suddenly a very deep gut feeling that something was different. It occurred when looking at Earth and seeing its blue and white planet floating here and knowing it was orbiting the sun, seeing that sun, seeing it set in the background of the very deep black and velvety cosmos. Seeing–rather knowing for sure–that there was a purposefulness of flow of energy, of time, of space in the cosmos — that it was beyond man's rational ability to understand, that suddenly there was a nonrational way of understanding that had been beyond my previous experience.

> There seems to be more to the universe than random, chaotic purposeless movement of a collection of molecular particles.

> On the trip home, gazing through 240,000 miles of space toward the stars and the planet from which I had come, I suddenly experienced the universe as intelligent, loving, harmonious.

In Einstein's *The World As I See It*:

> The most beautiful experience we can have is the mysterious. It is the fundamental emotion which stands at the cradle of true art and true science....I am satisfied with the mystery of the eternity of life and with the awareness and a glimpse of the marvelous structure of the existing world, together with the devoted striv-

ing to comprehend a portion, be it ever so tiny, of the Reason that manifests itself in nature.

And in Aczel's work *God's Equation*:

> In January 1998, the way we perceive the universe changed forever. Astronomers found evidence that the cosmos is expanding at an ever-greater rate. Soon as the new findings were announced, cosmologists from all over the world rushed to explain the underlying phenomenon. The most promising theory that scientists could come up with was one that Albert Einstein had proposed eight decades earlier and quickly retracted, calling it his greatest blunder.

Every year, new developments prove the accuracy of Einstein's theories. But if the cosmologists' new assessments are correct, then Einstein was right even when he thought he was wrong.

BIBLIOGRAPHY

Aczel, Amir D. *God's Equation: Einstein, Relativity, And The Expanding Universe*. New York, Four Walls Eight Windows, 1999. Quotation from Preface pp IX.

Barrow, John D and Frank Tipler. *The Anthropic Cosmological Principle*. Oxford University Press, 1986.

Bergman, Peter G. *The Riddle Of Gravitation*, New York, Charles Scribner's Sons, 1968.

Bortle, John E. *The Remarkable Case of Comet Lovejoy*. Sky and Telescope Vol 125, No 5, May 2012: pp 36-41.

Cattermole, Peter and Patrick Moore. *The Story Of The Earth*. London, Oregon Press, 1985.

Danielson, Dennis Richard. (ed.) *The Book Of The Cosmos*. Cambridge, Perseus, 2000.

Davies, Paul. *The Eerie Silence*. UK, Penguin, 2010.

Davies, Paul. *The Goldilocks Enigma*. UK, Penguin, 2006.

Delsemme, Armand. *Our Cosmic Origins*. Cambridge, Cambridge University Press, 1998.

Einstein, Albert. *The World As I See It*. New York, Crown, 1954: Quotation pp 9. Originally published in *Forum and Century*, Vol. 84, the thirteenth in the Forum series, *Living Philosophies*.

Ferris, Timothy (ed.) *The World Treasury of Physics, Astronomy and Mathematics*. Toronto, Little Brown, 1991.

Fortey, Richard. *Life*, New York, Random House, 1997. Quotation pp 141.

Gingerich, Owen. *The Book Nobody Read*, New York, Penguin, 2004.

Greenberg, Richard. *Unmasking Europa*, New York, Copernicus Books, 2008.

Grinspoon, David. "Life On Saturn's Moons?" *Sky And Telescope.* Vol 124, No 6, December 2012: pp18.

Hart, Michael H. *The 100. A Ranking of the Most Influential Persons In History.* New York, Citadel Press, 1978.

Hawking, Stephen and Leonard Mlodinow. *The Grand Design*, New York, Bantam, 2010.

Hawking,Stephen (ed.) *On The Shoulders Of Giants.* Philadelphia, Running Press, 2004.

Hecht, Jeff. "Evolving Planet." *Sky And Telescope.* Vol 120, No 2, August 2010: pp 20-26.

Hibbert, Christopher. *The Story Of England*, London, Phaidon Press, 1992.

Hinckley, John. *Star Names Their Lore and Meaning.* New York, Dover, 1963. First published by G.E. Stechert in 1899 as *Star Names and Their Meaning.*

Hirschfeld, Alan W. *Parallax The Race To Measure the Cosmos.* New York, Holt, 2001.

James, C. Renee and William Sheehan. "Neptune Comes Full Circle." *Sky And Telescope.* Vol 122, No 1, July 2011: pp 28-34.

Jofré, Paula and Payel Das. "The Evolution Of Spiral Galaxies." *Astronomy and Geophysics.* Vol 58, Issue 5, October 2017: pp 5.13- 5.17.

Kragh, Helge "Big Bang:The Etymology Of A Name. *Astronomy and Geophysics.* Vol 54, Issue 2 April 2013: pp 2.28-2.30.

Krauss, Lawrence M. *A Universe From Nothing.* New York, Simon and Schuster, 2012.

Kuhn, Thomas S. *The Copernican Revolution*, New York, Random House, 1957.

Lorentz, Ralph and Jacqueline Mitton. *Titan Unveiled.* New Jersey, Princeton University Press, 2008.

Mitchell, Dr Edgar with Dwight Williams, *The Way Of The Explorer.* New York, G.P Putnam's Sons, 1996.

Munitz, Milton K. *Cosmic Understanding.* New Jersey, Princeton University Press, 1986.

Munitz, Milton K. *The Question Of Reality.* New Jersey, Princeton University Press, 1990.

Newton, Sir Isaac. *Opticks*, New York. Dover, 1952. Based on the 4th Edition, London, 1730.

Pannekoek, Anton. *A History Of Astronomy*. London, George Allen & Unwin, 1961.

Penrose, Roger. *The Large, The Small And The Human Mind*. Cambridge University Press, 1997.

Penrose, Roger. *Cycles Of Time*. New York. Vintage Books, 2012.

Pettegree, Andrew. *The Book In The Renaissance*. Yale University Press, 2010.

Reymo, Chet. *Skeptics And True Believers. The Exhilarating Connection Between Science And Religion*. New York, Walker And Company. 1998.

Roberts, J.M. *A History Of Europe*, New York, Penguin Press, 1997.

Sagan, Carl and Iosef Schlovski. *Intelligent Life In The Universe*. San Francisco, Holden-Day, 1966.

Seielstad, George. *At The Heart Of The Web. The Inevitable Genesis Of Intelligent Life*. Orlando, Harcourt, 1989.

Singh, Simon. *Big Bang*, New York, Harper Collins, 2004.

Tarnus, Richard. *Cosmos And Psyche*. New York, Penguin, 2006.

Walker, James G., *Earth History: The Several Ages Of The Earth*. Boston, Jones & Bartlett, 1986.

Wilson, Robert. *Astronomy Through The Ages*. London, Taylor And Harris, 1997.

Ward, Peter and Donald Brownlee, *Rare Earth*. New York, Copernicus, 2000.

Weinberg, Steven. *Dreams Of A Final Theory*, New York, Random House, 1992.

Zwart, Simon F Portegies. *The Lost Siblings Of The Sun*, Astronomical Institute Anton Pannekoek, University of Amsterdam, March 2009.

INDEX

A

Absolute zero, 4, 95, 113, 114
ALH 84001, 187
Anaximander, 56
Angular momentum, 89, 177-179, 185-187, 189
Anthropic Principle, 153
Apollo missions, 23-27, 39, 40, 48, 89, 93, 94, 97, 102, 103, 148, 207
Aristarchus, 96, 175
Aristotle, 45, 64, 80
Asimov, Isaac, 168
Augustine of Hippo, 64

B

Bede, 61
Berkeley, George, 150
Big Bang, 11-14, 60, 63-65, 69, 71, 78, 115, 117-120, 123-125, 128, 129, 131, 150, 151, 153-155, 157-160, 166, 168-170, 195, 196, 205
Big Crunch, 65, 157
Big Freeze, 158
Big Rip, 158
Black holes, 16, 66, 130, 131, 134, 135, 137, 138, 140, 143, 153, 157, 158, 175
Bow Tie or Boomerang nebula, 114
Bradley, James, 70
Brahe, Tycho, 44, 70
Bruno, Giordano , 45, 146, 148

C

Caesar, Julius, 57
Cassini, Giovanni Domenico, 22, 32
Clarke, Arthur C., 145, 148, 168
Cosmological constant, 15, 119, 156, 158
Copernicus, Nicolas, 43, 44, 60, 91, 148, 176, 203, 205
Copernican Principle, 70, 148, 149
Crab Nebula, 133
Cygnus X-1, 138-140

D

Dark energy, 10, 12, 15, 16, 63, 68, 119, 126, 128, 130, 154-158
Darwin, Charles, 86, 109, 162, 166, 203-205
Dawn Mission, 16, 30, 43, 56, 89, 105
Democritus, 195, 196
Demosthenes, 67
Descartes, René, 49, 66, 176
Dicke, Robert, 13
DNA, 41, 42, 109, 113, 193, 194, 196, 198, 204, 206

E

Edgeworth-Kuiper Belt (see Kuiper Belt objects)
Einstein, Albert, 2, 14, 15, 43, 58, 60, 66-69, 72, 73, 116, 119, 135, 207, 208
Eris, 32

Ptolemy, 45
Pulsars, 132, 202
Pythagoras, 52

Q

Quasars, 122, 130, 131, 155

R

R136A1 star, 4, 5
Rare Earth Hypothesis, 163
Rømer, Ole, 66

S

Sagan, Carl, 51
Saturn, 6, 22, 25, 32, 36, 84, 89, 93, 101, 164, 181, 182, 184, 186
Sedna, 33
Shmaonov, T. A., 13
Sitter, Willem de, 13
Socrates, 167
Solar system, 5-7, 9, 22, 28, 29, 32, 34, 36-38, 52, 56, 70, 71, 78, 81, 84, 85, 89, 94-96, 98, 101, 131, 138, 143-145, 149, 150, 152, 164-166, 168, 172, 175, 176, 178-180, 183, 184, 186, 187, 196, 198
Spinoza, Baruch de, 112
String theory, 63, 69
Supernovae, 4, 7, 12, 114, 115, 123, 130-133, 136, 139, 140, 143, 145, 171, 175, 179, 180
Soyuz, 26, 27
Sputnik, 16, 17, 20

T

Teng Mu, 197
Theophilus, 64
Trans Neptunian Objects, 32
Tsiolkovsky, Konstantin, 19, 27, 28, 92, 103
Tunguska event, 40, 147

U

Uranus, 6, 22, 28, 32, 34, 36, 95, 96, 168, 182

V

V2 rocket, 18, 19
Venus, 6, 21, 22, 29, 84, 93, 102, 144, 172, 182, 187, 192
Von Braun, Werner, 18, 19
Voskhod, 24-26
Vostok, 21, 24, 25
Voyager missions, 22, 28, 36, 37, 51, 85, 192

W

Wells, H G , 1, 43, 60, 177
Weinburg, Steven, 41

Z

Zwicky, Fritz, 20, 130, 132

Printed in the United States
By Bookmasters